DEMOCRATIZING TECHNOLOGY

Democratizing Technology

Andrew Feenberg's Critical Theory of Technology

Edited and with an introduction by
Tyler Veak

STATE UNIVERSITY OF NEW YORK PRESS

Published by
State University of New York Press, Albany

Printed in the United States of America

For information, address State University of New York Press,
194 Washington Avenue, Suite 305, Albany, NY 12210-2384

Production by Kelli Williams
Marketing by Anne M. Valentine

Library of Congress Cataloging-in-Publication Data

Democratizing technology : Andrew Feenberg's critical theory of technology / editor, Tyler J. Veak.
p. cm.
Includes bibliographical references and index.
ISBN-13: 978-0-7914-6917-0 (hardcover : alk. paper)
ISBN-10: 0-7914-6917-4 (hardcover : alk. paper)
ISBN-13: 978-0-7914-6918-7 (pbk. : alk. paper)
ISBN-10: 0-7914-6918-2 (hardcover : alk. paper) 1. Technology—Social aspects.
I. Veak, Tyler J., 1963-
T14.5.D448 2006
303.48'3—dc22 2005037169

10 9 8 7 6 5 4 3 2 1

Contents

Acknowledgments vii

Introduction ix

PART 1: THEORETICAL ASSUMPTIONS OF A CRITICAL THEORY OF TECHNOLOGY

Chapter 1. Rethinking Modernity as the Construction of Technological Systems 3
David J. Stump

Chapter 2. The Posthuman Challenge to Andrew Feenberg 19
Simon Cooper

Chapter 3. An Ecofeminist Response 37
Trish Glazebrook

Chapter 4. What's Wrong with Being a Technological Essentialist? A Response to Feenberg 53
Iain Thomson

Chapter 5. From Critical Theory to Pragmatism: Feenberg's Progress 71
Larry A. Hickman

PART 2: THE POLITICS OF TECHNOLOGICAL TRANSFORMATION

Chapter 6. Democracy and Technology 85
Gerald Doppelt

Chapter 7. Feenberg and the Reform of Technology 101
Albert Borgmann

Chapter 8. Commodification and Secondary Rationalization 112
Paul B. Thompson

Chapter 9. Democratic Technology, Population, and Environmental Change 136
Andrew Light

Chapter 10. Technological Malleability and the Social Reconstruction of Technologies 153
Edward J. Woodhouse

Replies to Critics 175
Andrew Feenberg

Contributors 211

Index 217

Acknowledgments

The idea for this volume was first conceived at a Society for Philosophy and Technology conference in 1999, where I presented a paper critiquing Feenberg's latest work, *Questioning Technology*. Feenberg responded; and from this engagement we began an ongoing dialogue that ultimately resulted in this volume. When I initiated this project I had no idea that it would take seven years to complete. Putting together an edited volume requires the cooperation of multiple parties without whom this volume would not have been possible.

I wish to thank all of the contributors for their insightful chapters and for their patience in waiting for this publication, and especially to Andrew for taking the time to work with me on this project. I am also grateful to Jane Bunker at SUNY Press who was willing to take on such a project from an unknown scholar, and to Kelli Williams for her patience in leading me through the maze of bringing a manuscript to publication. I would also like to thank my good friend and fellow colleague, Piyush Mathur, for his superbly prepared index.

Lastly, I want to thank my family for their unwavering support in all my endeavors. . .

Tyler Veak
Lynchburg, Virginia

Introduction

Given the interconnections between particular technologies and local/global problems such as war, poverty, environmental destruction, disease, and increasing economic disparity, the importance of formulating a theory of technological transformation seems paramount. To this end, Andrew Feenberg offers one of the most fully developed theories on the politics of technological transformation to date.[1] His critical theory of technology is, therefore, a significant point of dialogue for further research, hence the reason for this volume.

Feenberg argues that "there are ways of rationalizing society that democratize rather than centralize control."[2] He reasons that if modernity as we know it is established through a process of rationalization, then alternative rationalizations are necessary in order to create alternative modernities. According to Feenberg, the current modernity is characterized by a particular rationality—a technical code—and that this rationality has been embodied in the technological designs of modern society. Democratizing technology means expanding technological design to include alternative interests and values.

Background and Context of Feenberg's Work

Some understanding of the context of Feenberg's work is necessary to fully appreciate the contributions in this volume. Feenberg, a student of Herbert Marcuse, draws most heavily from the Frankfurt School[3] tradition to formulate his critical theory of technology. Like his Frankfurt School predecessors, Feenberg's work is largely a response to, or continuation, of Max Weber's theory of modernity. Weber claimed that the process of modernization fueled by capitalism's emphasis on "formal rationality" necessarily led to a differentiation between technological and social spheres.[4] In short, the progress of modernity was achieved at the expense of moving away from the personal (substantive) relations of traditional societies to the impersonal (formal) relations of modern society. According to Weber, capitalism adopts formal rationality to achieve increased control, the end of which is total bureaucracy—the "iron cage."[5] Driven by human needs, capitalism attempts to maximize production through formal rationality. The more rationalized a system becomes the more it produces. End of story. There cannot be a normative assessment of such a value-neutral system.

The degree to which Frankfurt Institute members[6] borrowed from the work of Weber cannot be overemphasized. Like Weber, Theodore Adorno and Max Horkheimer held that in the context of capitalism "useful" translates into economically valuable.[7] Although the drive for domination was not new to the Enlightenment, it was the employment of new tools of domination—science and technology—that made the Enlightenment program unique. Technology as instrumental reason was one of the primary means of instilling this domination.[8] Reason was therefore not the road to emancipation hailed by Enlightenment thinkers, but a new method of shackling humanity.

Herbert Marcuse elaborated these themes in his characterization of a "one-dimensional society."[9] According to Marcuse, technology has been co-opted through a political choice to establish the present form—"technological rationality." He argues that reason manifested through technology serves as an instrument of domination: "Today, domination perpetuates itself and extends itself not only through technology, but as technology. . . . Technological rationality thus protects rather than cancels the legitimacy of domination."[10]

Marcuse claims that the discourse of rationality established and maintained by the mass media, essentially negates all opposition. The discourse loses its play, dialogue, mediation, and consequently its ability to create new alternatives. It becomes tautologous, and in doing so it contains those aspects of the discourse that could offer an alternative to the status quo. It has, in Marcuse's words, created a one-dimensional society: "There is only one dimension, and it is everywhere and in all forms."[11] Class consciousness attained in the context of late capitalism is necessarily a false consciousness, "and this false consciousness has become embodied in the prevailing technical apparatus, which in turn reproduces it."[12] In the one-dimensional society, whatever "is" is right, and wrong is only that which is antithetical to the whole—the one-totalizing, all-pervading, self-legitimating discourse of truth.

Although Marcuse was skeptical about the prospects for transformation, he did believe that it was possible. His proposal for liberation involves three elements: a critique of the existing technological consciousness, a new subject, or agent, and a new technique of pacification. He claims that the contradictions created by the hyperrational, technological society open the door for critique.

This is not a belief in the inherent breakdown of economic conditions, as scientific Marxists argued. Marcuse, instead, argues that late capitalism generates enough negative externalities to make us question the rational nature of the given technological society. This is where critical theory comes in; it "strives to define the irrational character of the established rationality."[13]

Marcuse claims that the choice of an alternative technology is limited by: (1) the "stuff" of Nature, however defined, as it confronts the subjective interpreter, and (2) the form of interpretation in a given cultural historical context.[14] In other words, both the inherited technological (material) context and the given technological consciousness represent important constraints on technological choice. Alternative historical projects must be at least imaginable for the possibility of "an ingression of liberty into historical necessity."[15] Alternatives, however, are difficult to achieve because of the hegemony of the few who have control over the productive process.

The issue of agency is crucial; someone must be capable of acting. Although not particularly hopeful, Marcuse argues that the best opportunity for alternatives lies with the excluded, or marginalized. That is, the revolutionary standpoint rests with the outsiders and outcasts, such as people of color, the persecuted, and the unemployed. All they need is the consciousness to act.[16]

Although it is a given that Nature must be pacified for human existence, this mastery can be either repressive, or liberatory. Marcuse suggests a new "technique of pacification" to replace the established technology of domination. To accomplish this, Marcuse claims that technology must be redefined as an "art of life."[17] He sees the development of the aesthetic dimension as central to this liberation project.[18] Beauty must become a form of freedom. "Rather than being the handmaiden of the established apparatus, beautifying its business and its misery, art would become a technique for destroying this business and this misery."[19] And with the emancipation of Nature comes the simultaneous emancipation of human senses. We are thereby liberated to experience gratification from Nature in a multitude of new ways. Humans, however, must begin to appreciate Nature for its own sake—"a subject with which to live in a common human universe."[20]

Jürgen Habermas, a second-generation critical theorist, fundamentally opposes Marcuse's call for a new technology. According to Habermas, technology is essentially the unburdening of needs that are rooted in human nature through purposive-rational action (i.e., work).[21] Technology is who we are—the innate faculty of "purposive-rational action" enabled humans to control their environment and set themselves apart from the rest of Nature. For Habermas, suggesting a new technology is as absurd as suggesting a new human species. Liberation cannot be achieved by transforming technology because technology simply cannot be altered.

Habermas admits that the technological rationalization of society is problematic but that technology itself is not the cause. The source of the trouble lies in the tension between the spheres of work and communication. The domination of work, or purposive-rational action, over traditional forms of communication is a defining feature of the modern period.[22] Habermas claims that Marcuse is misguided in his call for a revision of the human-Nature relationship with an alternative mediating technology. Marcuse, according to Habermas, fails to distinguish between two different types of action. The human-Nature relationship is necessarily governed by "purposive-rational action." What Marcuse is advocating, on the other hand, is actually "communicative action."[23]

Like Weber (and the Frankfurt School), Habermas sees technology as instrumental rationality; specifically, he defines technology as "scientifically rationalized control of objectified process."[24] However, his approach to resolving the problem is somewhat different: "*Rationalization at the level of the institutional framework* can occur only in the medium of symbolic interaction itself, that is, through *removing restrictions on communication*."[25] Emancipation is therefore achieved not through a new technology as Marcuse advocates but by limiting purposive-rational subsystems through "new zones of conflict."[26] The central problem is not technology itself because, again, technology cannot be changed.

Feenberg's Critical Theory of Technology

With Weber as a backdrop, Feenberg combines the insights of Marcuse and Habermas to arrive at his own "critical theory of technology." He believes Marcuse was correct to argue that technology is to a large extent socially shaped and that the form of technology is a political choice. The problem with Marcuse's perspective is that he does not acknowledge the contingency within the technologically dominated one-dimensional society. For Marcuse, it is either all or nothing—a technique of domination or a technique of liberation. This is what leads Marcuse to argue that transformation must come from "outside" the system; those within the one-dimensional society are simply too constrained to act. Feenberg rejects this appeal to outsiders as the basis for transforming society. He argues that the goal is "not to destroy the system by which we are enframed but to alter its direction of development through a new kind of technological politics."[27] The aim, in other words, is to steer the system from within through subtle hybridizations not mass revolution.

With modifications, Feenberg employs Habermas's model of a democratic speech community as the means for liberating technological design choice from hegemonic constraints.[28] Whereas Habermas argues for the exclusion of technological rationality from the lifeworld of communicative action, Feenberg brings rationality into Habermas's vision of a democratic community to arrive at his suggested "democratic rationality." Contra Habermas, this process of transforming technology must take place within the social.[29] The possibility exists to choose rationally more liberating technological designs that further the various interests of the community of actors. As Feenberg states, "There are ways of rationalizing society that democratize rather than centralize control."[30]

Although Feenberg draws heavily from Marcuse and Habermas to formulate his critical theory of technology, he attempts to eliminate their essentialist base.[31] Feenberg claims that the tendency to essentialize is due primarily to an overemphasis on the meta-level of culture. In the case of Habermas, this resulted from his sidelining of the technology-society relationship to focus on language and communication. This is even more obvious in the case of the Frankfurt Schoolers, such as Marcuse, who framed technology as an autonomous, rationalizing force acting hand-in-hand with capitalism to produce agentless workers/citizens of a one-dimensional society. Their infatuation with Weber's concept of rationalization, combined with the philosophical shift in Marxism,[32] predisposed the Frankfurt School to focus on consciousness, or ideology.

Feenberg's central point is that technology can only be misconstrued as an autonomous-rationalizing force if the contingency evidenced at the micro-level of design is ignored. Although technology frequently appears to have an essence because it is viewed ahistorically, there is actually no "essence" of technology. Feenberg argues that scholarly interpretations of the social construction of technology (SCOT)[33] establish convincingly that technical design can only be defined contextually and locally by the particular technology-society relationship.[34] There is a significant degree of contingency, difference, or, "interpretive flexibility" in a society's relationship with particular technologies.

While SCOT reveals insight on the technology-society relationship, Feenberg rightly points out its deficiencies. SCOT is too narrowly focused on the development of particular technological artifacts, or systems.[35] Wanting to include all elements in the analysis is understandable; however, SCOT takes the concept of symmetry[36] too far in an attempt to level the playing field. SCOT ignores the larger issue of how particular design choices are made over other choices, which, as Feenberg argues, is an inherently political question. He claims that since technology can never be removed from a context, it can never be neutral.[37] Technological design is inherently political; consequently, the observed constraint on design choice is not some essence of technology but evidence of the hegemonic control of the design process by privileged actors.

There is, however, an obvious tension between the *contingency* observed at the level of design choice, and the *constraints* placed on design by the larger cultural-political milieu. Feenberg characterizes this tension as the "ambivalence" of technology, which he conveys in the following two principles:

1. *Conservation of hierarchy:* social hierarchy can generally be preserved and reproduced as new technology is introduced. This principle explains the extraordinary continuity of power in advanced capitalist societies over the last several generations, made possible by technocratic strategies of modernization despite enormous technical changes.
2. *Democratic rationalization:* new technology can also be used to undermine the existing social hierarchy or to force it to meet needs it has ignored. This principle explains the technical initiatives that often accompany the structural reforms pursued by union, environmental, and other social movements.[38]

Feenberg admits that advanced societies concretize power through technologically mediated organizations that prevent their citizens from meaningful political participation. Focusing on this aspect of culture led the Frankfurt School to characterize technology as an autonomous, rationalizing force. The problem is that they ignored the existence of the second principle of "democratic rationalization." Feenberg believes that democratic rationalization can overthrow this entrenched power "from 'within,' by individuals immediately engaged in technically mediated activities and able to actualize ambivalent potentialities suppressed by the prevailing technological rationality."[39]

He claims that "strategic" actors are able to realize their particular biases in to the technological designs. These biases stem from

> aspects of technological regimes which can best be interpreted as direct reflections of significant social values in the "technical code" of the technology. *Technical codes define the objects in strictly technical terms in accordance with the social meaning it has acquired.* These codes are usually invisible because, like culture itself, they appear self-evident.[40]

According to Feenberg, control over design choice is not always economically motivated, as Marxists frequently argue. That is, the utilitarian efficiency of the market is not always the motivating factor. Frequently, the aim is to either de-skill workers, or for management to maintain operational autonomy.[41] A centralized-hierarchical power structure is perpetuated because technological designs (codes) are intentionally chosen to maintain operational autonomy. Feenberg therefore admits that although technocratic power is foundationless and contingent, it nevertheless has a "unidirectional tendency."[42]

Despite this fact, Feenberg believes that it is possible for "tactical" actors to subvert the established technical code through their own democratic rationalizations. Feenberg provides examples of what he considers successful democratic rationalizations of technology, such as the struggle over the Internet and AIDS activists' reform of the FDA drug approval process. Although the Internet was originally designed for the transmission of data, interpretive flexibility enabled a multitude of users to shape the Internet for their own uses.[43] In the case of AIDS treatment, activists collectively challenged traditional medicine's technocratic view of treatment. Activists forced a dialogue with research scientists and the FDA. In the end, they successfully altered the entrenched government bureaucracy to gain access to experimental medicines, which in turn led to significant advancements in treating AIDS.[44]

Instrumentalization Theory

The tensions between the principles of the "conservation of hierarchy" and "democratic rationalization" can also be discussed in terms of an *analytic* distinction between primary and secondary instrumentalization. Feenberg differentiates "primary instrumentalization" in which the functional, reifying aspects of technology are emphasized, from "secondary instrumentalization" where objects are constituted into their particular social contexts. Primary instrumentalization is therefore analogous to Weber's (and the Frankfurt School's) discussion of technology as formal rationality. Habermas, in agreement with Weber, sees differentiation (i.e., between technological and social spheres) as the unavoidable consequence of technological modernization. Feenberg in contrast argues that this differentiation is more apparent than real. That is, looking at technology at the meta-level makes it appear that technology has a differentiating effect.

He characterizes primary instrumentalization with four moments. Technology has the effect of *decontextualizing* entities from their original context. The qualities of objects are *reduced* to quantifiable terms so that they can be easily controlled with the established laws of science and technology. Those in control seek to *position* themselves strategically in order to more easily exert their power. All of this assumes a degree of *autonomization*, or distancing, between those in control and the objects being controlled.[45]

Feenberg suggests that primary instrumentalization and the process of differentiation can be overcome not through containment, as Habermas argues, but through a process of subversive, or secondary, instrumentalization. The overall thrust of this level is to *realize* elements that have been decontextualized through the process of primary instrumentalization. Feenberg offers four secondary moments to counter the reifying moments of primary instrumentalization. Actors must seek to *systematize* elements that have been decontextualized.[46]

Figure 1. Feenberg's Theory of Instrumentalization[47]

Differentiation

←

Primary instrumentalization	**Secondary instrumentalization**
Decontextualization	Systematization
Reduction	Mediation
Autonomy	Vocation
Positioning	Initiative

→

Realization

In response to the reductionism of primary instrumentalization, actors can re-instill objects with secondary qualities. Aesthetic and ethical *mediations* are added to technological objects when recontextualized. Autonomization can be overcome through *vocation*, or the actual way in which users engage technologies. Although the distancing effect of technology is real, actors can make choices in the way in which they actually employ technologies. Finally, tactical actors can exert their *initiative* to counter the positioning of strategic actors attempting to control them through technology.[48]

Technological Consciousness

In order to move toward realization, the hegemony constraining design choice must be exposed. What is needed, according Feenberg, is a theory of cultural change: "A new culture is needed to shift patterns of investment and consumption and to open up the imagination to technical advances that transform the horizon of economic action."[49] He draws on a number of intellectual traditions—hermeneutics from Heidegger, cultural theory from Foucault and Baudrillard, and critical theory—to reveal how the interests of certain actors achieve and maintain control of the design choice process.

Feenberg argues that the essentialist view of technology as inherently differentiating is actually a product of a reified technological consciousness. He compares this consciousness to Marx's discussion of commodity fetishism[50] in which commodities are reified and treated as actually existing, autonomous entities:

> The fetishistic perception of technology similarly masks its relational character: it appears as a non-social instantiation of pure technical rationality rather than as a node in a social network. Essentialism theorizes this form and not the reality of technology.[51]

The technological consciousness reifies the split between primary (functional) and secondary (all other) qualities. Feenberg, however, argues that this division follows from modern society's exaggerated emphasis on functionality. In its extreme form this begins to look much like modernity's hyperrationalization of society through capitalism's emphasis on utilitarian efficiency. It is possible to describe technological objects using both functional and social language, which is what Feenberg means by contrasting the moments of primary and secondary instrumentalization. But again, it must be understood that this is an analytic distinction and that only part of the secondary instrumentalization can be considered social.[52]

Organization of Volume

Contributors to this volume both critique and build on Feenberg's efforts to establish a theory of technological reform. Although there is no clear way of segregating the articles, they do fall roughly into two groups. Chapters in Part 1 center loosely on Feenberg's theoretical assumptions. The chapters in Part 2, on the other hand, focus more on the politics of technological reform. Regardless, the division is not simply between theory and practice as both parts contain elements of each.

In the first chapter, David Stump takes issue with Feenberg's use of the social construction of technology (SCOT). Stump claims that SCOT is itself essentialist in that it makes a priori assumptions about the design process. In short, SCOT is too narrowly focused on the social, or political aspects of technological design. Placing too much emphasis on the social aspects is what leads to Feenberg's overly optimistic picture of technological transformation. Taking all factors into consideration, one may conclude that many entrenched technologies are extremely difficult to alter.

In addition, Stump argues that Feenberg's critical theory of technology is also essentialist because it relies on a notion of "technology in general." The problem is that there is a tension between providing a normative approach to technological transformation and avoiding essentialism. Stump, however, suggests that it is possible to adopt an anti-essentialist approach by making generalizations from the historical analysis of particular technologies and then applying these generalizations normatively.

In chapter 2, Simon Cooper challenges Feenberg's theoretical approach with a discussion of the "ontological contradictions" created by emerging biotechnologies. These technologies have the potential of simultaneously enhancing our lives, and destroying our contexts of meaning. According to Cooper, biotechnologies are creating a "posthuman" future categorically different than any suggested by Feenberg's "alternative modernities." Cooper argues that Feenberg's theory of transformation is too dependant on established cultural reference points (e.g., democratic norms), which will radically change in a posthuman world. Cooper argues that we must ask broader questions to effectively engage these emerging technologies.

Trish Glazebrook (chapter 3) approaches Feenberg's work as a Heideggerian ecofeminist. She is excited about the prospects of linking her work with the specifics of Feenberg's politics of technological transformation. More constructive, rather than critical, Glazebrook establishes a number of connections between ecofeminism and Feenberg's politics of technology. First, technology and women's bodies are both sites of political struggle and resistance. Second, Feenberg's constructivist view of technology resonates with the prevailing feminist view of gender as constructed. Third, Feenberg's concept of "subversive rationalization" offers a powerful reply to those who persist in labeling feminists as "anti-rationalists."

Glazebrook, however, objects to Feenberg's critique of Heidegger. She argues that Feenberg is more Heideggarian than he admits, and that he also displays essentialist tendencies in his discussion of instrumentalization. Glazebrook argues that the Heideggerian perspective on technology as "ways of thinking" is still useful. In addition to our actual practice, we need a theoretical/ideological shift in the way we engage technology.

In chapter 4, Iain Thomson takes issue with Feenberg's framing of Heidegger as a technological essentialist. Thomson claims that Feenberg's critique of Heidegger's essentialism can be broken into three types: ahistoricism, substantivism, and one-dimensionalism, and then proceeds to show how none of these critiques actually apply to Heidegger's views. Thomson hopes that his analysis will "vindicate Heidegger's ground-breaking ontological approach to the philosophy of technology," and "help to orient the approach of future philosophers of technology to one of its central theoretical controversies" (i.e., essentialism).

In the fifth chapter, Larry Hickman spins a pragmatist's web around Feenberg's work and claims that John Dewey anticipated much of his theory of technology by several decades. Most notably, Dewey's theory of technoscience, like Feenberg's, was anti-essentialist, constructivist, and democratic in nature. But in addition, Hickman claims that Dewey's theory is more developed than Feenberg's because he offers a detailed theory of democracy which Feenberg does not.

Part 2 begins with a chapter by Gerald Doppelt who sees Feenberg's project as two distinct parts—the first deals with the demystification of old theories of technology, and the second a new normative, critical theory of technology. Doppelt is convinced by Feenberg's critique of essentialist theories of technology; however, he does not believe that Feenberg supplies the necessary "ethical resources" for his "democratic rationalization of technology." He agrees with Feenberg's claim that the potential exists for marginalized interests to shape the design process—Feenberg's examples establish this. However, Doppelt asserts that the notion of "participant interests" is inadequate to supply a conception of what a "democratized" technology actually is, or should look like. In other words, Doppelt believes that it is not sufficient to simply open up the design process to other interests; there must be some way of asserting *which* interests are better, or more democratic than others. Doppelt argues that Feenberg's critical theory must be supplemented with liberal-democratic concepts such as "entitlements" and "equality" to achieve this grounding.

In chapter 7, Albert Borgmann uses Feenberg as a platform to discuss the criteria of technological reform. He believes that Feenberg's theory meets the criteria of feasibility, cultural depth, scope, and substantive content. Borgmann points out, however, that Feenberg's own examples indicate that there is a limit to the democratization of technology. That is, the costs involved in extending secondary instrumentalization all the way down can be excessive. In addition, Borgmann believes that Feenberg fails to address two important hindrances to reform: (1) the fact that the majority of people choose affluence over autonomy, and (2) the enormous cost in terms of time and money that is required to effectively engage technology. Borgmann offers examples to illustrate these points and suggests avenues for future research that could lead to overcoming these obstacles.

Paul B. Thompson (chapter 8) suggests ways of extending Feenberg's theory of instrumentalization by borrowing from the field of institutional economics and the philosophy of Albert Borgmann. Thompson makes the important distinction between "structural" commodification and "technological" commodification. Structural commodification involves changes in the rules, laws, or social customs associated with a particular technology. In these instances, the actual physical technology itself is not altered.

In the case of technological commodification, the technological artifact itself is altered. The four parameters of alienability, excludability, rivalry, and standardization are actually built into the technological design. Thompson offers the invention of sound recording as an example of this type of commodification. He explains that "the advantages of the new vocabulary are an increased capacity to map the complexities of commodification and decommodification, and in a clearer way to express how technology and technological innovation affects those processes." Thompson concludes by illustrating the merits of his suggested vocabulary through a reexamination Feenberg's examples of AIDS activism and the Internet.

In the penultimate chapter, Andrew Light addresses Feenberg's views on environmentalism. He begins by criticizing Feenberg's assessment of the Erlich-Commoner debate over population,[53] arguing that this debate does not provide an accurate terrain of contemporary environmentalism. Light believes that environmentalism, and its relationship with technology, is far more complex than Feenberg insinuates. The issue of justice, according to Light, offers a better lens through which to analyze contemporary environmentalism.

Light also takes issue with Feenberg's suggested appropriation of environmentalism for the democratic reform of technology. Feenberg asserts that environmentalist's values can provide the catalyst and guide for the democratic reform of non-sustainable technologies. Light claims that Feenberg creates a black box out of environmentalism and uses it as a means to his end. As an alternative, Light wants to emphasize environmental management *practices*, and offers restoration ecology as an illustration. Restoration ecology is a technology that has the potential of facilitating the development of values conducive to the long-term establishment of sustainable communities. In other words, it is the practice or process that is important and not simply the values going into the process as Feenberg seems to suggest.

Ned Woodhouse, in the final chapter, largely agrees with Feenberg's project. His criticisms center around Feenberg's approach and the examples he uses to make his argument. Like Feenberg, Woodhouse agrees that technology is inherently contingent and malleable. However, Woodhouse believes that Feenberg's examples (e.g., AIDS activism) do not go the heart of contemporary society. Woodhouse suggests that movements such as "green chemistry" offer greater potential for change.

In terms of approach to reform, Woodhouse argues that Feenberg's methodology is too large in scope. He claims that Feenberg's project implies a wholesale replacement of our political and economic system (i.e., with socialism). Woodhouse, on the other hand, wants to take a more piecemeal approach by focusing on particular "elements of technological governance." He offers corporate executive officer incentive programs as one possibility for accomplishing this goal.

Feenberg concludes the volume with a reply to the contributors. His response is insightful because he attempts to situate their arguments within the broader context of his theory and the problems confronting technological reform in general. Clearly, there are gaps between the contributors' views and Feenberg's. The aim of this volume is to facilitate further dialogue and fill in the sketch provided by Feenberg's foundational work. Hopefully, others will be inspired to continue the important project of constructing an "alternative modernity."

Notes

1. Feenberg has written a trilogy of books and numerous articles on the subject. See Andrew Feenberg, *Critical Theory of Technology* (New York: Oxford University Press, 1991); Feenberg, *Alternative Modernity: The Technical Turn in Philosophy and Social Theory* (Los Angeles: University of California Press, 1995); Feenberg, *Questioning Technology* (London and New York: Routledge. 1999).
2. Feenberg, *Questioning Technology*, 76.
3. What has come to be known as the Frankfurt School of critical theory originated in the Institute for Social Research (Institute) as part of the Frankfurt Institute in 1923.
4. Max Weber, *Economy and Society: An Outline of Interpretive Sociology*, 3 vols., ed. Guenther Roth and Claus Wittich (New York: Bedminster Press, 1968).
5. Herbert Marcuse, *Negations: Essays in Critical Theory*, trans. Jeremy J. Shapiro (Boston: Beacon Press, 1968), 203.
6. Key members of the Institute included Max Horkheimer, Herbert Marcuse, Theodore Adorno, Friedrich Pollock, Erich Fromm, and Leo Lowenthal. See Douglas Kellner, *Critical Theory, Marxism, and Modernity* (Baltimore: John Hopkins University Press, 1989), 12.
7. Theodore Adorno and Max Horkheimer, *Dialectic of Enlightenment*, trans. J. Cummings (New York: Herder and Herder, 1972 [1944]).
8. David Held, *Introduction to Critical Theory: Horkheimer to Habermas* (Berkeley: University of California Press, 1980), 170.
9. Herbert Marcuse, *One-Dimensional Man: Studies in the Ideology of Advanced Industrial Society* (Boston: Beacon Press, 1964), 154–55.
10. Ibid., 158–59.
11. Ibid., 11.
12. Ibid., 145.
13. Ibid., 227.
14. Ibid., 218.
15. Ibid., 221.
16. Ibid., 256.
17. Ibid., 338.
18. Herbert Marcuse, *The Aesthetic Dimension: Toward a Critique of Marxist Aesthetics* (Boston: Beacon Press, 1978).
19. Marcuse, *One-Dimensional Man*, 239.
20. Herbert Marcuse, *Counterrevolution and Revolt* (Boston: Beacon Press, 1972), 60.
21. Jürgen Habermas, *Toward a Rational Society: Student Protest, Science, and Politics*, trans. Jeremy J. Shapiro (Boston: Beacon Press, 1968).
22. Ibid., 96.
23. Ibid., 88.
24. Ibid., 57.
25. Ibid., 118.
26. Ibid., 120.
27. Feenberg, *Alternative Modernity*, 35.
28. Jürgen Habermas, *The Theory of Communicative Action*, 2 vols., trans. Thomas McCarthy (Boston: Beacon Press, 1984–1987).

29. Feenberg, *Alternative Modernity*, 81.
30. Feenberg, *Questioning Technology*, 76.
31. According to Feenberg, essentialist philosophies of technology originated with Heidegger and were further developed by the Frankfurt Schoolers (*Questioning Technology*).
32. The failure of a proletariat-led revolution initiated a shift in Marxist thought in the early twentieth century. In short, this involved a moving away from Scientific Marxism, or economism, to a more philosophically oriented theory. Karl Korsch and Georg Lukács are largely credited with initiating this shift.
33. Feenberg broadly conceives SCOT to include social constructivists, contextualist historians of technology, and actor-network theorists. There have been a number of edited volumes on the SCOT since the current rage began in the early 1980s. See for example, Donald A. MacKenzie, and Judy Wajcman, eds., *The Social Shaping of Technology: How the Refrigerator Got its Hum* (Philadelphia: Open University Press, 1985); Weibe Bijke, Thomas Hughes, and Trevor Pinch, eds., *The Social Construction of Technological Systems* (Cambridge: MIT Press, 1987); Weiber Bijker and John Law, eds., *Shaping Technology/Building Society: Studies in Sociotechnical Change* (Cambridge: MIT Press, 1992).
34. Feenberg, *Questioning Technology*, 78–83.
35. Ibid., 11.
36. The concept of symmetry has its origins in the "Strong Program" of the sociology of scientific knowledge. The idea is that the analyst must suspend truth or falsity in order to give all perspectives a fair analysis. See David Bloor, *Knowledge and Social Imagery* (London; Boston: Routledge, 1973).
37. Feenberg, *Questioning Technology*, 213.
38. Ibid., 76.
39. Ibid., 105.
40. Ibid., 88.
41. Feenberg, *Alternative Modernity*, 87.
42. Ibid., 92.
43. Feenberg, *Questioning Technology*, 126.
44. Feenberg, *Alternative Modernity*, ch. 5.
45. Feenberg, *Questioning Technology*, 203–204.
46. Feenberg claims that this is roughly analogous to Bruno Latour's discussion of "enrolling" actors into networks. Latour, for example, uses Pasteur's development of germ theory to illustrate how system builders enroll actors into their networks. See Bruno Latour, *The Pasteurization of France* (Cambridge: Harvard University Press, 1988).
47. Used by permission of author (Feenberg, *Questioning Techology*, 221).
48. Ibid., 205–207.
49. Ibid., 98.
50. The commodity form mystifies the productive relations that go into the product and thereby hides the exploitative labor relations that produce the commodity. See Karl Marx, *Capital: A Critical Analysis of Capitalist Production*, vol. 1, ed. Frederick Engels (New York: International Publishers,1967[1867]), 77.
51. Feenberg, *Questioning Technology*, 211.
52. See Feenberg's discussion of "Instrumentalization Theory" in this volume.
53. Discussed in chapter 3 of *Questioning Technology*.

PART 1

Theoretical Assumptions of a Critical Theory of Technology

CHAPTER ONE

DAVID J. STUMP

Rethinking Modernity as the Construction of Technological Systems

Introduction

Andrew Feenberg has carved out a unique and philosophically productive position in the philosophy of technology that is informed by both the essentialist philosophy of technology developed by Heidegger and the Frankfurt School, and by various social historical accounts of science and technology such as those developed Weibe Bijker, T. P. Hughes, and Bruno Latour.[1] These two schools of thought are not easily integrated, for although they share technology as a major subject of their study, their central understandings of the relationship between technology and society contradict one another. This chapter will focus on what philosophers of technology should learn from the social historians of technology and will argue that a full acceptance of the view of the social historians of technology will improve the position of critiques of technology. Feenberg's merging of the philosophy of technology and science studies gives rise to questions about the concept of modernity, since modernity is characterized by the development of technology.

Max Weber's analysis of modernity is crucial to understanding the debate between various Frankfurt School writers on the philosophy of technology because all of them adopted Weber's view that modernity is defined by differentiation—the fragmentation of the world into autonomous "spheres of value" such as the political sphere, the economic sphere, etc. Since each of these spheres operates independently and according to its own aims, methods, and logic, each can advance unimpeded by "external influences." The autonomous development of science and technology in the industrial age and of the economy in capitalist free markets is emblematic of modernity and is the prime examples of its progress. However, this same autonomy of the spheres implies that our technological processes are "meaningless," since they stand outside of any unifying worldview and, even worse, are out of our control due precisely to their autonomy. In the hands of experts who follow only the internal logic of technological development, technology is a dominating force rather than a tool that we can control and use for our benefit, since it is beyond the reach of political action or other forces that are independent of its own logic. Even advocates of modernity such as Habermas think that because technology is natural, it is in some sense unquestionable. One cannot be against science or technology since it is part of human nature to seek knowledge and to use things as means to human ends.[2] Feenberg rejects this technological determinism, referring to social constructivist accounts of technology to make his case.

Social constructivist accounts question the distinctive autonomy or differentiation of spheres in modernity by pointing out with historical case studies that there is always a mixture of technical, political, economic, and other social concerns in the development of technology, not just technological imperatives. The autonomy of technology seems to be nothing but a myth from this point of view, since social and political factors always influence decisions made in technology and science. Indeed, the central aim of the social constructivist histories of science and technology is to prove that there are always social and political dimensions of technology. The philosophy of technology of both antitechnological writers such as Heidegger and pro-Enlightenment writers such as Habermas are undercut by social histories of science and technology, Heidegger's because his essentialism fails to account for either the historical development of technology or the intersection of technology with economic, social, and political concerns.[3] Habermas's fix for the Enlightenment project, the governance of the lifeworld by a communicative rationality that is independent of and sheltered from technical rationality, is also shown to be untenable by the social constructivist histories of science and technology. Habermas accepts the autonomy of technical rationality in a limited role running the system, while maintaining the democratic aspirations of the Enlightenment by underscoring the role of communicative reason in the lifeworld and in providing legitimization for the political order, but the very distinctions that are supposed to save the modern picture have been called into question, as has Weber's definition of modernity itself.[4]

Far from limiting philosophical critique of technology, however, social constructivist accounts can be understood as proof that technology is open to criticism and to change. With examples developed in several of his works, Feenberg argues convincingly that consumers, students, and workers have already had important impacts on the development of technology, shaping it through social negotiation or by finding unintended uses for technology. Just as Monsieur Jourdain of Molière's play "The Bourgeois Gentlemen" is surprised and pleased to learn that he has been speaking prose all his life, readers used to the pessimism of essentialist philosophy of technology may be surprised and pleased, when reading Feenberg, to discover that people have been successfully struggling against technology all along and have forced significant changes in the way technology has developed and been applied. This recognition of the place for social and political shaping of technology that has been shown in social and historical accounts of the development of technology is the most innovative and powerful aspect of Feenberg's philosophy of technology. However, there are several questions that are raised by the merging of these two incompatible viewpoints: What can we say about Weber's influential and helpful definition of modernity in the light of its apparent inconsistency with the social constructivist view of technology? How much freedom of choice and political control can the public really exercise over the development of science and technology? Do we need to maintain that there is some kind of essence of technology in order to be able to analyze and criticize it philosophically? Extending Feenberg's use of social and historical studies of science and technology, I will argue that one can maintain Weber's definition of modernity as differentiation within the framework of the social historical accounts of technology and that while a fully anti-essentialist point of view provides more resources for the political and social grounding of technology than Feenberg recognizes, it also correctly implies that freedom of choice and the possibility of political action in a technological society depend on the specifics of the case involved, and indeed thus may be extremely limited in some circumstances. Trying to maintain a position intermediate to those of essentialist philosophers of technology and anti-essentialist social constructivists is difficult, but let us hope that Feenberg will respond like the philosophical master who says, after Monsieur Jourdain expresses his sympathy for the beating that the philosopher received from Jourdain's other instructors: "'Tis nothing at all. A philosopher knows how to take things and I'll compose a satire against them, in the manner of Juvenal, which will tear them to shreds. Let that pass. What do you wish to learn?"[5]

The Social Construction of Knowledge, Science, and Technology.

When Feenberg appropriates social constructivist accounts of technology, he ignores the varieties of positions in science studies so as not to divert himself or his readers with internecine battles, but there is one that should not be overlooked. The social constructivist version of science and technology studies is not only unjustified by the historical record, it is also essentialist in ways that should be troubling to Feenberg, and it is even flagrantly inconsistent with his own description of the development of technology. There is no reason for Feenberg to associate himself with social construction when there are much more persuasive and informative views that provide a better basis for the philosophy of technology and will provide far less misleading accounts of the development of technology.

Feenberg is correct when he says that the simplest argument for the social construction of scientific knowledge has always been the underdetermination argument (the Duhem-Quine thesis), which supposedly shows that technology and science are never objectively justified since there are always alternative theories that can be as well justified as the theory under consideration.[6] However, there are very good arguments in the literature to show that underdetermination is much more limited than commonly thought.[7] An exchange between David Bloor and Bruno Latour nicely illustrates the issues at stake between social constructivists and their critics among social historians of technology.[8] Bloor wants to use social science as a tool to explain why scientists and engineers adopt the beliefs that they do, while Latour, quite vocal in his opposition to social construction for a long time, says that we cannot *explain* such belief formation at all. Despite his protestations to the contrary, Bloor ends up reducing science and technology to a social phenomenon. Since he claims to have shown by the underdetermination argument that logic and empirical evidence are never sufficient to explain why a particular theory or method is adopted, any explanation that Bloor can offer must be in terms of social conditions or agreement.[9] However, choices that face the developers of a new technology are at least as underdetermined by social factors as by internal ones. As Latour has emphasized, social factors alone do not explain technological change.[10] Latour's motivation for adding nonhuman agents to his actor-network theory is precisely to avoid reducing science to a social phenomenon. Even more problematic is the fact that the strict distinction between the content of a theory and the social conditions surrounding it cannot be maintained under the current historiographic standards. The concrete setting of technological development always includes social or political, as well as technical or immanent concerns, and all of these need to be included in an account of technological development.[11]

Latour has followed Bijker, Hughes, and other historians of technology in developing the idea that there are multiple constraints on the development of technology and science, some social, some economic, some intrinsic to the science involved, etc. Whether social context or technical constraints offer the most appropriate explanation of some aspect of the development of a particular technology is, of course, heavily dependent on what questions are being asked and will vary according to the specific issues that developers of specific technologies face, but the central claim of Latour and others is that both society and the content of science are required for a full understanding of such development.[12] Feenberg's account is actually closer to the social historians of technology than to the social constructivists, since his discussion is full of "constraint talk," as Andy Pickering would put it.[13] For example, Feenberg discusses the "positioning" of technologies, giving examples of all of the laws that different technical spheres posit in order to circumscribe their actions: "The laws of combustion rule over the automobile's engine as the laws of the market govern the investor."[14] There are constraints in each of Weber's spheres of value. "These 'laws' might be scientific, but they also might be something like deep public attitudes, which advertisers must work with rather than against, or in the case of management, assumptions about the psychology of workers."[15] This seems obviously correct, but also quite contrary to social construction, at least as developed by Bloor and the so-called strong programme. In a purely logical, or an abstract philosophical sense, developers of technology can do anything that they want, but as a practical matter, they are constrained by social norms, by the economic market, by their own ingenuity, by the disagreement of relevant experts, and by the difficult processes of getting their machinery to work (getting nature to agree, so to speak).

For Feenberg, the central point to be appropriated from historians of technology is that people are always involved in technological development. While it is true that both the social constructivists and the social historians of technology emphasize this human involvement, the social constructivists' position is incompatible with Feenberg's own accounts of the development of technology, as I have noted. Social constructivism is also essentialist, adopting two of the three specific elements of essentialism that Feenberg rejects in his discussion of that doctrine.[16] First of all, by basing their claims on the underdetermination thesis, the social constructivist's account of the development of technology becomes essentially ahistorical. As many have noted, social constructivists do not need detailed historical studies to make their case for the underdetermination of theories by evidence, since they are fundamentally using an a priori philosophical argument that applies to everything, independent of the time, place, subject matter, and social and political conditions. In their eagerness to prove their point, the social constructivists have moved far away from their supposed empirical and historical sociology, and indeed have shown little more than an abstract tautology. A weather report that guaranteed its accuracy by predicting that there will be some kind of weather tomorrow, either wet or dry, warm or cold, would not be informative. The mere logical possibility of creating alternative technologies is likewise uninformative, since political action requires real alternatives. Genuine historical studies can be very illuminating concerning the options and choices that scientists actually made; underdetermination arguments cannot. The social constructivist position is also one-dimensional in a way that should trouble Feenberg. Since science and technology are really just another type of politics or social interaction, according to social constructivists, the many roles that modern material technology or technological thinking play in culture are ignored, indeed even eliminated. Regarding a third element of essentialism, substantivism is directly contradicted by social historical accounts that reject the claim that science and technology are independent forces imposing values on us. Feenberg, of course, uses the results of these social constructivist accounts to show that technology is more open to political challenge than essentialist philosophers of technology thought. It is here that social constructivist accounts comes into conflict with Weber's influential analysis of modernity, as Feenberg himself recognizes. "But if technology is so profoundly embedded in the social, differentiation must be far less thoroughgoing than essentialist theories of modernity assume."[17]

The Essence of Technology

Just as Feenberg does not entirely adopt social constructivism, neither does he entirely adopt anti-essentialism, as is indicated by several passages in the book where he concurs with essentialist critiques of technology. However, Feenberg's attempt to keep the general analytical framework that the essentialist account of technology makes available, while at the same time rejecting essentialism and, indeed, showing forcefully how a fixed essence would make it impossible to change technology and thus get in the way of the positive program he develops for democratizing it, introduces a tension into his work. In chapter 9 of *Questioning Technology*, Feenberg develops a positive program for the analysis of technology incorporating both general and concrete analysis, thus synthesizing philosophical essentialist and concrete, particularistic accounts of technology under the banner of social construction.

The differences between the general properties of technology and its specific applications show up in Feenberg's account as "primary and secondary instrumentalizations," which describe the general characteristics of technology and a schema for the characteristics of technology that appear when technologies are actually implemented, respectively. However, to begin an analysis of technology in general such as the primary instrumentalizations require, one must assume that one can meaningfully discuss technology in general—that there is a single process common to technological development, an essence of technology in at least some schematic form, or a single technological relation to the world.[18] The four elements of primary instrumentalization are decontextualization, reductionism, autonomization, and positioning. Heidegger and other essentialists have often said that technology tears objects from their meaningful context and shows them to us as a resource. The tree loses its place in the forest and is seen as lumber, but the lumber would not exist qua lumber without carpenters and the housing market, so objects are always recontextualized in a new social setting. Feenberg also says that technology reduces things to their primary qualities, that is, their former social relations are typically ignored when they are put into a scientific or technological context. However, decontextualization is never literally true, even if it might be a useful analytical term. Likewise, though technology is designed to create autonomous decision making, Feenberg's examples of successful democratic resistance to technology show precisely the limits of this autonomy. Secondary instrumentalization, subtitled realization, relates to how technology is put into concrete practice and again has four aspects: systematization, mediation, vocation, and initiative. As Feenberg emphasizes, the central question of technology is how much genuine choice and control we have when technology is put into practice. There is no alternative to making detailed concrete studies of the development and implementation of a technology, in order to answer this question.

Modernity without Essence

According to Weber, the Enlightenment project of making individuals and society rational has led to each "sphere of value" running by its own rules and aims. Therefore, modern politics is separated from religion, modern scientific knowledge follows its own methods and studies everything without being concerned with consequences, and a modern free market follows the laws of supply and demand without "artificial" limits on growth, profit or loss, etc. Weber's basic idea is that progress has occurred as people moved away from personal connections to impersonal ones. Early societies were based on family and tribal relationships that were then replaced by the "ethic of brotherliness," while in modern society we have completely abstract, impersonal relationships, at least in the public spheres, and an increasing awareness of the internal autonomous logic of each sphere.

> [T]he rationalization and the conscious sublimation of man's relations to the various spheres of values, external and internal, as well as religious and secular, have then pressed towards making conscious the internal and lawful autonomy of the individual spheres; . . . This results quite generally from the development of inner- and other-worldly values towards rationality, towards conscious endeavor, and towards sublimation by knowledge.[19]

Weber's idea of differentiation as essential to modernity is not necessarily incompatible with the social studies of science and technology. The idea of a totally value-free technology *is* a myth, as social historians have argued, but technology, the economy, and politics really have become autonomous in the modern era, relative to their position in the premodern eras. When social constructivists say that technology is political, they do not mean that it literally includes democratic institutions, but rather that human interactions are hidden and distorted in technology. Feenberg correctly emphasizes developing a new type of politics in response to the development of technology in order to respond to the impact of technology on society.

Feenberg is also on the right track in seeing the development of technology and modernity as historical. What is the difference between an essentialist understanding of technology and a historical one, if both agree that a historically coherent set of practices can legitimately be called modern technology? First, a historical conception does not support teleology—the development of technology is radically contingent. Second, there is no essence, or single formula for technological development, but rather major differences in how specific technologies are developed and implemented. Nevertheless, we can still find evidence for the Weberian analysis of modernity in anti-essentialist, local studies of science and technology. Technology is not one thing and it is not automatically autonomous and independent of human interactions, but it can *become* independent and autonomous. The large technological systems that define modern culture and are emblematic of modern rationality are out of our control, but not because they are outside of human culture, but rather because they are larger and stronger than most of the social mechanisms that we might use to influence them. It takes a vast social movement or a major catastrophe to change a fully developed large-scale technology.

The development of entrenched technological systems can be seen clearly in Hughes's landmark study of Edison's creation of an electrical power network of which his famous light bulb is but one small part. While Hughes is at pains to argue that Edison's having the right science is not enough to explain the success of his technology, it is also clear that scientific and technical success is required along with political, financial, and marketing success. So, when writing a history of a technological development, many heterogeneous elements will have to come into play. The result is that neither social factors nor technological innovation can be said to drive change; technology shapes society and society shapes technology.

> In a sense, electrical power systems, like so much other technology, are both causes and effects of social change. . . . This book is not simply a history of the external factors that shape technology, nor is it only a history of the internal dynamics of technology; it is a history of technology and society.[20]

The early electric lighting demonstration projects at Pearl Street in New York, Holborn Viaduct in London, and Friedrichstrasse and Markgrafenstrasse in Berlin marked the birth of a power grid that today reaches almost every home in the industrialized world.

A key concept in Hughes's analysis of the development of technology is that of a system: "[In its broader uses] 'system' then means interacting components of different kinds, such as the technical and the institutional, as well as different values; such a system is neither centrally controlled nor directed towards a clearly defined goal."[21] Large complex technological systems embody the modern world and are unique to it. As artifacts, they exhibit the physical, intellectual, and symbolic resources of our culture, but they also shape our culture, because the physical space in which we live and the objects we use affect us by circumscribing our possible actions. Systems such as electrical power networks reassemble the differentiated spheres to create a concrete whole, while at the same time establishing themselves as an autonomous power that is very difficult to challenge. Hughes' concept of a system is the key to understanding Weber's conception of modernity in terms of the new social history of technology. It is not science and technology per se that are autonomous, it is rather established technological systems that become autonomous, because they become harder and harder to oppose or to influence as they grow. Hughes recognizes four stages of the development of a system—invention and development, technology transfer, system growth, and substantial momentum.[22] As the name suggests, it is in the last stage that a technological system takes on a life of its own, and it is precisely in this stage that Hughes finds that technological systems start shaping society. Substantial momentum is also the stage of development in which technologies are appropriated by market forces that are out of the hands of the inventor, ultimately taken over by financiers.[23] Even though we praise scientific and technical genius, twentieth-century scientific and technological systems are collective efforts for which no one individual can be responsible. Locating autonomy in the technological system, rather than in technology in general, shows that the autonomy of technology can exist even if social and political factors are always involved in the development of technology. Large technical systems of all sorts really are out of our control, since there is no way for a small group to make the needed changes. Americans are very dependent on their automobiles, for example, and many things would have to change before and during a transition to a less automobile-dependent way of life. The autonomy of any particular technological system is not guaranteed, however, since each technical system needs to grow to the point where it has substantial momentum before it becomes a force that is autonomous.

Issues for Future Research

The philosophy of technology must change substantially to accommodate the lessons of the social histories of technology written in the last twenty-five years. In order to facilitate the integration of these two fields, four topics that require further reflection will be discussed here: the local nature of technological development, the contingent but real autonomy of technology, the control of technological systems, and the normativity of the philosophy of technology. Feenberg is well aware of these issues and addresses them by maintaining elements of essentialism in order to maintain an adequate analysis of modernity and to support normative claims. Here I will argue that an anti-essentialist account of technology can successfully address these concerns, if the philosophical implications of the social historical accounts of technology are clarified.

Accepting current standards of historical scholarship, social histories study the heterogeneous influences on the development of technology in a local setting, showing how scientific knowledge, material laboratory practice, market forces, social values, and politics are all integrated in the creation of a particular technological system. Technology is thus the paradigm case of local, contingent, temporal, and embodied knowledge, while philosophy has traditionally been concerned with universal, necessary, eternal, and disembodied knowledge. Essentialist philosophies of technology are manifestly atemporal, while the aims and methods, the institutional organization, the content of scientific theories, and the material engineering practice of technological development all change through time. Adhering to current historiographical standards requires eschewing general philosophical arguments, looking instead at the development of each technology to see how choices were made and to see what alternatives were left undeveloped. Embracing a historical method could seem to eviscerate any possible critique of technology, since social histories are limited to studies of the development of particular technologies at a particular time and are therefore unable to make general claims that describe an essence of technology. If studies of individual technological systems cannot be generalized, a critical analysis of the general meaning of technology will be lost.

In a sense, this problem of generality is merely a recapitulation of the problem of induction. Empirical historical studies will necessarily be limited to particular cases and the conclusions drawn will not be general. However, as Callon and Latour have emphasized, while the lesson to learn from the demise of intellectual history is that connections between historical players and ideas must be shown in the local setting, there is no intrinsic limit to the evidence that can be gathered nor on the generality of the claims that can be made.[24] While empirical studies will never achieve the certainty and necessity claimed by a priori philosophical method, they can be as general as the evidence allows. To ask for anything more seems unreasonable, just as Hume's demand that induction be justified according to a deductive standard has seemed unreasonable to many philosophers. If one can find a way to make connections between disparate events, one will have a general analysis, even while meeting the requirements of an empirical, local, concrete case study. Elements of the essentialist critique of technology could therefore be justified by empirical evidence. For example, the demand for efficiency could be seen as a common standard that is found in multiple empirical studies of modern society to be applied to an increasing number of areas over time, rather than as an essential feature of technical rationality.

Alternatives to current technologies and social structures will in some cases be extremely limited. Feenberg appeals to social construction in order to show that technology is always open to social criticism. While he is certainly right to reject technological determinism, there are some technological systems that are autonomous and so firmly embedded in our culture that they appear everywhere, but at the same time remain invisible, which is exactly why they are so powerful. Foucault makes this point in *Discipline and Punish:* "How could the prison not be immediately accepted when, by locking up, retraining and rendering docile, it merely reproduces, with a little more emphasis, all the mechanisms that are to be found in the social body? The prison is like a rather disciplined barracks, a strict school, a dark workshop, but not qualitatively different."[25] Electrical power networks, transportation systems, and communications systems are taken for granted as available resources—we hardly notice them until they fail. When they do fail, however, we see not only to what extent these systems touch every aspect of our lives, we also see the extent to which an overwhelming majority of the public wants these systems. To say that it is possible to challenge any technological development is one thing, but to actually change an entrenched system is another. The issue of how political philosophy can be a guide to action, offering real possibilities instead of merely logical ones, is not merely a practical question, nor a minor difficulty, but rather the major question facing those who would challenge an entrenched technological system.

Technological systems that have achieved substantial momentum in Hughes's sense are far more autonomous and powerful than any current human social movement. Action is always possible, but there are fundamentally different constraints on the types of actions that must be taken in order to change social structures and technological systems. Even though they are human constructs and embedded in social relations, technological systems can become autonomous and very difficult to challenge. I suggest that we look at the technological systems as quasi-objects—natural/social hybrids, in Latour's vocabulary. The autonomy of highly developed technological systems is just as real, in a practical sense, as electricity or microbes. Latour accepts the reality of specific microbes before and after their discovery by Pasteur, comparing their existence to the existence of an electrical network and explicitly citing Hughes. It is rather odd that Latour then rejects the autonomy of technological systems, claiming that we have never achieved the separation of science from society that was promised in the Enlightenment, when these very systems serve as a model of his hybrids.[26] Modern society has created technological systems that are autonomous in a practical sense. Autonomous hybrids are not equal, since Latour's blurring of the distinction between nature and society does not leave us with a multitude of undifferentiated hybrids, but rather with partially ordered sets of hybrids, some of which are more autonomous than others. In this sense the distinction between technological systems and society can be maintained.

If technological systems really do develop autonomy and an unstoppable momentum, we need to reconsider our normative stance toward technology and those who benefit or suffer from it. Technology may seem to develop autonomously, so that no individuals can be held responsible for it. There are too many independent forces in the heterogeneous mix of factors that can affect the development of a technology for any individual to claim credit or blame. The contingency of the development of technology emphasized in social historical accounts further limits the responsibility of individuals, since rational planning has less to do with technological success than formerly thought. The development of technology does not follow predictable paths, let alone the inexorable one imagined by the technological determinist. Individuals cannot be blamed for what they cannot control or what they cannot foresee, so it might look like it is impossible to blame anyone for the development of technology, if we adopt an anti-essentialist position. In order to assign blame to those in charge of technology, Feenberg appeals to essentialism:

> There are, as essentialists argue, technological masters who relate through rational planning to a world reduced to raw materials. But ordinary people do not resemble the efficiency oriented system planners who pepper the pages of technology critique.[27]

Individuals and classes do have very different positions in modern technological society, but no one is "in charge," a point made most forcefully by Foucault.[28] Foucault uses Bentham's ideal prison, the Panopticon, as an emblem of efficient power. The power relations are so integral to the architecture of the prison that prisoners are controlled even when there is no one in the guard tower. Those in the guard tower are merely employees who are evaluated and judged just like the prisoners. They are treated less harshly but controlled more insidiously, since the guards are convinced that they are free. Even great corporations have to please their shareholders and individual consumers, as well as enlist the cooperation of their suppliers and distributors, their machines and workers. The actions of corporations are highly constrained by the global and local economic systems. However, the presence of technological masters is not necessary to criticize modern technological society, since even if no one is in charge, some actors still benefit from technological development more than others, which is enough to raise issues of social justice. If we look at concrete historical studies of specific technological changes, there will be winners and losers (consider the average CEO salary in America) and those who have tremendous advantages or disadvantages throughout their lives. Feenberg and I agree that the philosophy of technology should support critical stances toward the inequalities that exist in society, but we disagree as to whether remnants of essentialism are needed to provide philosophical grounding for a social or political movement.[29]

The historical method could also seem to eviscerate any possible critique of technology, if social histories may seem to be limited to descriptive accounts of the development of technology and are unable to make normative claims. First, the fact/value distinction seems to imply that empirical studies alone can never lead to normative claims, because inferring value statements from facts, that is, from descriptive empirical discoveries, commits the naturalistic fallacy. However, social histories of science and technology have shown that we cannot understand science if we only look at it as pure knowledge. The institutional structure of science, its relation to the greater society, etc., are all relevant to understanding science and technology, not just the resulting knowledge. Furthermore, pure knowledge may not be as value neutral as it is typically assumed to be. For example, some "mere facts" express significant value by radically changing our self-conception. The widespread reaction to the publication of tentative evidence for the existence of life on Mars is a contemporary example of this phenomenon,[30] the Copernican and Darwinian revolutions serve as others.[31] Second, we see another version of the problem of generality mentioned above. It is often thought that a local, nontheoretical analysis will be inadequate to ground a political movement. Applying Callon and Latour's idea for grounding general claims with empirical studies to the grounding of normative stances toward technology, we will find that there is no theoretical limit to the critical force that an empirical study of technology will have. The political force that empirical studies possess is a practical, concrete question, not an abstract issue of grounding. To make one's position known and to have an effect, one must muster allies, be read, cited, and so on, and make a practical difference.[32]

The four issues raised above initiate further discussion and reflection on aspects of Feenberg's fruitful approach to the philosophy of technology. The far more difficult project that lies ahead is actually changing the current social structure and technological practice. In recognizing the significance of social histories of technology and in reconceptualizing the philosophy of technology, Feenberg has performed an enormous service. Let us hope that many will join him, that the field will flourish and make an impact on social and political action, as well as thought.

Notes

1. Bijker, Hughes et al., *The Social Construction of Technological Systems* (Cambridge: MIT Press 1989).
2. Jürgen Habermas, "Technology and Science as 'Ideology,'" in *Toward a Rational Society:Student Protest, Science, and Politics*. Trans. Jeremy J. Shapiro (Boston: Beacon Press, 1970), 81–122, 87.
3. For a discussion of Heidegger, see Thomson this volume and Andrew Feenberg, "The Ontic and the Ontological in Heidegger's Philosophy of Technology: Response to Thomson," *Inquiry* 43, no. 4(2000): 445–450.
4. See Habermas 1970, 91 ff. Habermas begins with a distinction between work and [social] interaction. Work is governed by empirical results, while interactions are governed by social norms that are conventional and ideally arrived at by consensus. Having isolated two very different kinds of rules and norms, Habermas suggests that there are two very different forms of the rationalization of society. Therefore, technical progress can be acknowledged without threatening the goals of the Enlightenment. However, the problem with the modern world is that once the traditional unified worldview that served to legitimate the social order broke down, nothing replaced it. Attempts to legitimate the social order with scientific/technical reasons or with individual subjective reasons both fail, according to Habermas. Furthermore, neither of these types of legitimation are opened to public debate, showing how typical attempts to legitimate modern social structures move away from the democratic ideals of the Enlightenment.

5. Molière, "The Middle Class Gentleman," 29, trans. H. Briffault (Woodbury, NY: Barron's Educational Series, Inc., original edition, *Le Bourgeois Gentilhomme*, 1670, Act II, scene 4).
6. Andrew Feenberg, *Alternative Modernity: The Technical Turn in Philosophy and Social Theory* (Berkeley: University of California Press, 1995), 3.
7. See Peter Galison, *How Experiments End* (Chicago: University of Chicago Press, 1987); David J. Stump, "From Epistemology and Metaphysics to Concrete Connections," in *Disunity of Science: Boundaries, Contexts, and Power*, ed. Peter Galison and David J. Stump (Stanford: Stanford University Press, 1996), 255–286; Elliott Sober, "Testability," *Proceedings and Addresses of the American Philosophical Association* 73, no. 2 (1999): 47–76; Elliott Sober, "Quine's Two Dogmas," *Proceedings of the Aristotelian Society*, supplementary vol. 74 (2000): 237–80. Neither Duhem nor Quine advocated relativism. Duhem says that we always have two choices—conservatism, holding on to our existing theories and changing auxiliary assumptions to accommodate the new facts, or radicalism, replacing the old theory with a fundamentally different one that accounts for the new facts and the old ones too. Duhem says both choices are rational despite the fact that there is no formal method by which to make a decision, because a scientist can act as an impartial judge: "The day arrives when good sense comes out so clearly in favor of one of the two sides that the other side gives up the struggle even though pure logic would not forbid its continuation." Pierre Duhem, *The Aim and Structure of Physical Theory* (Princeton: Princeton University Press, 1954[1905]).
8. See David Bloor, "Anti-Latour," *Studies in History and Philosophy of Science* 30A, no. 1 (1999): 81–112; David Bloor, "Reply to Bruno Latour," *Studies in History and Philosophy of Science* 30A, no. 1 (1999): 131–136; Bruno Latour, "For David Bloor . . . And Beyond: A Reply to David Bloor's Anti-Latour," *Studies in History and Philosophy of Science* 30A, no. 1 (1999): 113–129. For a recent comprehensive survey, see John H. Zammito, *A Nice Derangement of Epistemes: Post Positivism in the Study of Science from Quine to Latour* (Chicago: University of Chicago Press, 2004).
9. Stump, "Concrete Connections," 262–263.
10. Bruno Latour, *The Pasteurization of France* (Cambridge: Harvard University Press, 1988); Bruno Latour, "One More Turn after the Social Turn," in *Social Dimensions of Science*, ed. E. McMullin (Notre Dame: University of Notre Dame Press, 1992), 272–294.
11. See Stump, "Concrete Connections." The distinction between the content of a theory and the social conditions is very similar to the distinction between the context of justification and the context of discovery first expressed by Reichenbach. Although widely challenged, this distinction is still invoked in the defense of science against attacks by social constructivists. My approach is very different, in that I use the lack of such a distinction to challenge social construction, arguing that the social constructivist position cannot even be formulated without relying on such a distinction.
12. The same point can be seen clearly in Latour's own works on technology. See Bruno Latour, "Mixing Humans and Nonhumans Together: The Sociology of a Door-Closer," *Social Problems* 35(1988): 298–310, Bruno Latour, *Aramis or the Love of Technology* (Cambridge: Harvard University Press, 1996), and his account of Pasteur, where he specifically cites Hughes in order to explain that Pasteur neither simply discovered microbes, nor simply invented them, but

rather created a system of food production and public health that included microbes as an essential component ("Mixing Humans," 261–62, n.18, and n.24). Latour also, perhaps charitably, reads *Leviathian and the Air Pump* this way, taking the central point of Shapin and Schaffer's book to be that science and society develop together and mutually influence each other. Bruno Latour, *We Have Never Been Modern* (Cambridge: Harvard University Press, 1993).

13. Andrew Pickering, "Beyond Constraint: The Temporality of Practice and the Historicity of Knowledge," in *Scientific Practice: Theories and Stories of Doing Physics*, ed. Jed Z. Buchwald (Chicago: University of Chicago Press, 1995), 42–55.
14. Andrew Feenberg, *Questioning Technology* (New York: Routledge, 1999), 204.
15. Feenberg, private correspondence.
16. Feenberg, "The Ontic"; Thomson.
17. *Questioning Technology*, 210.
18. I have already raised some specific objections to Feenberg's account elsewhere and Feenberg has replied very satisfactorily, so here I merely outline his view before discussing an alternative account of the role of technology in modernity. See David J. Stump, "Socially Constructed Technology," *Inquiry* 43, no. 2 (2000): 1–8; and Andrew Feenberg, "Constructivism and Technology Critique: Reply to Critics," *Inquiry* 43, no. 2 (2000): 225–38.
19. Max Weber, "Religious Rejections of the World and Their Directions," *Max Weber: Essays in Sociology*, ed. H. Gerth and C. W. Mills (New York: Oxford University Press, 1946), 323–57, 328.
20. Thomas P. Hughes, *Networks of Power: Electrification in Western Society, 1880–1930* (Baltimore: Johns Hopkins University Press, 1983), 2.
21. Ibid., 6.
22. Ibid., 14–15.
23. Ibid., section 19.
24. Michel Callon and Bruno Latour, "Unscrewing the Big Leviathan: How Actors Macro-Structure Reality and How Sociologists Help Them to Do So," in *Advances in Social Theory and Methodology: Toward Integration of Micro- and Macro-Sociologies*, ed. Karin Knorr-Cetina and Aaron Cicourel (Boston: Routledge, 1981), 277–303.
25. Michel Foucault, *Discipline and Punish: The Birth of the Prison* (New York: Vintage Books, 1979), 233.
26. "Mixing Humans," 80 and 93.
27. *Questioning Technology*, x.
28. *Discipline and Punish*, 207.
29. "Constructivism," 231–32.
30. David S. McKay et al., "Search for Past Life on Mars: Possible Relic Biogenic Activity in Martian Meteorite Alh84001," *Science* 273, no. 5277 (August 16, 1996), 924–30.
31. Someone might reply that it is not the fact itself, but rather our reaction to it, that is the source of value, just as many have replied to John Searle's derivation of ought from is by arguing that he relies on the existence background practices that are the source of value. John R. Searle, "How to Derive 'Ought' from 'Is'," *Philosophical-Review* 73 (1964): 43–58. These background practices and social contexts are real, however, and completely intertwined with science and technology. They can be separated only in an abstract sense that hampers our understanding of technology.
32. "Concrete Connections," 285.

CHAPTER TWO

SIMON COOPER

The Posthuman Challenge to Andrew Feenberg

Introduction

Andrew Feenberg's work on technology has been motivated by a desire to reveal its contingent nature. To insist on contingency opens up the possibility of an "alternative modernity," a differentiated social and cultural framework that can engage productively with technology and not simply be subject to an unavoidable technological logic of rationalization and domination. One of the chief merits of Feenberg's work on technology lies in the way he has grappled with the legacy of modern philosophy and theorized how to say "yes" to technology. This "yes" is of course no simple or naïve affirmation of technology; rather, it is a critically informed recognition of the *ambivalence* of technology. Technology can reduce human freedom, especially in a world where capital remains the driving impulse, but technology can also be harnessed to serve more humane and democratic ends. This essay attempts to explore the degree to which Feenberg's strategies for securing an alternative modernity can succeed in the context of emerging technologies that move us closer toward the "posthuman."

Feenberg's approach to technology is complex and is the result of his attempt to harness the critical impulse from a variety of sources—Heidegger, the Frankfurt School, Haraway, and Latour—while avoiding their theoretical or political limitations. Feenberg attempts to strike a balance between technology's functional tendency toward reification and the human capacity for reflexive agency, which allows us to resist this tendency. As such, his work marks a vital contribution toward a politics of technology. Specifically, Feenberg's work contributes toward a post-Foucauldian politics of technology, albeit one that remains true to the larger project of making technologies more democratic and humane. I want to argue, however, that he has tipped the balance toward a pragmatic politics involving issues concerning technical design and outcomes, and that his attempt to rework modern substantive theory is limited in what can be stated about technology in the larger sense. (I also claim that something needs to be said in this context.) In his effort to avoid what he regards as the overdetermining and essentializing assumptions of substantive theorists of technology, Feenberg has perhaps cut himself off from insights into the broader context through which contemporary societies draw their values and meanings, and through which they might relate to any technological future. In proposing an alternative modernity, based upon cultivating the ambivalence of technologies, Feenberg has underplayed the larger shifts that have reconstructed the way in which we engage with the world. The consequence of this is that the values through which we might creatively harness this ambivalence have themselves been reframed. Examining the question of the "posthuman" enables us to explore this issue.

While Feenberg has dealt with most of the dominant theoretical approaches to technology, has applied his own theories to a variety of technologies, from on-line education to AIDS medicine, and has also drawn upon a variety of contexts through which to come to understand technology—from the culture of Japanese Go to modern spy films, he has not to date said a great deal about emerging technologies such as biotechnology. It may be that he has simply not got around to discussing biotechnology in any detail. It may be that he does not regard the "posthuman future" that such technologies propel us toward as categorically different from any other kind of future where human cultures intertwine with technological innovations, and that to think otherwise is to make precisely the kind of error he argues that substantive theories of technology have been making for decades. Thus, to pose such a "question concerning biotechnology" invites the criticism of essentialism, an inability to grant any capacity for agency, a lack of differentiation, and so on. It is likely that he regards biotechnology as being no different from other technologies, in the sense that their "challenges" can be met by a theoretical politics based around a commitment to democratic and humane use and design of technology.

I do not believe that this is ultimately the case. Indeed, the question of the posthuman provides a lens through which we can focus on the limits of Feenberg's approach to technology *in general*. Feenberg's two-level theory of instrumentalization allows for a "reflexive meta-technical practice,"[1] but this is a reflexivity that may not be adequate to the potentially transformative possibilities of technologies such as biotechnology. Biotechnologies promise the transcendence of many long-held social, biological, and cultural limits. The capacity for IVF, cloning, embryonic stem-cell research, or organ transplantation to enable liberation from one's embodied being carries with it a similar transcendence of the sociocultural frames that granted meaning to embodied ways of life. This contradictory form of liberation needs to be considered in more detail if we are to think about the radical prospects that the technosciences now offer us. In other words, a fully reflexive approach to technology needs to go beyond just the issue of the social and political control of technologies.

In relation to this one only has to look at how the question concerning biotechnology has split traditional political formations. Those on the Left who feel uneasy about biotechnology—either from a sense that it alters our fundamental humanness, or that it represents the final commodification of human life—often find themselves in the same camp as religious and political conservatives. Others on the materialist Left see such technologies in terms of emancipation and human progress, the issue of who has political control of technologies being the only substantive issue. The ongoing split on the Right between moral conservatives and economic libertarians is only exacerbated by the contemplation of a biotechnological future. Such fragmentation indicates the limits to which questions concerning biotechnology can be resolved along orthodox political lines.

The Question Concerning Biotechnology

Biotechnology is not simply a set of practices that can manipulate nature. After all, agricultural and medical practices have been doing this for decades. Instead, biotechnology encompasses contemporary technoscientific practices that enable the reconstitution of our social and cultural frames of reference—in particular our embodied sense of relatedness to the other—within a more abstract framework. Biotechnologies, as opposed to biological practices within modernity, move us toward an emerging ontological framework called the posthuman. The difference between modern and postmodern/posthuman biological practices can be registered through a comparison between modern and postmodern weaponry. Analytically, it is possible to distinguish weapons of "mass destruction" (nuclear, biological) from conventional weapons that can do a great deal of damage (such as firebombing from the air), but which are not predicated upon a reconstitution of the material realm. By contrast, in the case of both biotechnology and nuclear weapons a technoscientific intervention has been made into categories regarded as stable. Both rely on the technoscientific capacity to take hold of reality and to reconstitute it.[2] When speaking of technology, Feenberg rarely makes such a distinction. For him technology largely operates as an element of modernity and is always subject to political appropriation. The wider reconstructive potential of technology is downplayed.

This is not to say that Feenberg does not recognize the capacity for technology to create more abstract ways of being-in-the-world. It is implicit in his discussion of "primary instrumentalization"[3]—technology decontextualizes precisely through the fact that it can lift subjects and objects out of their prior frameworks. Yet the contradictory nature of this process is left largely unexplored. The more abstract ways through which we engage with the world—reaching out through media and information to transcend our more localized frames of reference has been an emancipatory process as much as a negative one. Yet this process cannot simply be analyzed through Feenberg's concept of "ambivalence," rather, it is something more like a process of *ontological contradiction*.[4] In other words, it is not simply that technology can enable or restrict human freedoms but that it alters the grounds through which our relation to those freedoms come into being.

Insofar as human needs and desires are carried within different historically constituted frames, it is necessary to consider whether technology is able to consummate these needs, or whether the reconstituting process it enables works to undermine the ground that historically sustained them. For instance, if globalization allows for a greater sense of connection with others across time and space, it also has the potential to erode the contexts that grant meaning to the connective process in the sense that the tangible presence of the other is no longer a defining dimension of the causal setting for interaction. Similarly, biotechnologies promise to enhance human life, but can undermine the contexts of embodiment and tangible relatedness which have granted meaning to life as an embodied subject. One can see the contradiction surrounding an early biotechnological process such as in vitro fertilization. If IVF technologies extend the possibilities of having a child they also undermine the sense of embodiment and relatedness that have historically and culturally framed what it means to have a child. The emancipation offered by reproductive technologies liberates the subject from the limitations of embodiment, but in doing so radically transcends the traditional liberal notion of the right to self-actuality. The cultural significance of childbirth is necessarily altered within this new technological framework. Perhaps more than any other technology, biotechnology articulates such contradictions.

Let me make it clear that the mere fact that technology allows for such reconstitution does not mean that it ought to be rejected. It is by focusing on the notion of ontological contradiction, however, that we can ask broader question about what social and cultural practices come to *mean* within the new context.

For the most part, however, the notion of contradiction has been sidelined. Attempts to critique biotechnology have generally fallen into two categories, both of which answer the question: "What are we protecting when we criticize biotechnology?" differently. The first is critical of biotechnology, and usually wants to reject it in toto because it violates nature. Biotechnology threatens to undermine basic human categories: our bodies (via organ transplants and biomedicine), our personalities (through neuropsychological drugs), our identities (through cloning), and so on. Often such critique comes from a conservative moral position.

The most prominent example of this position would be Francis Fukuyama's *Our Posthuman Future*, which argues that biotechnologies need to be strongly regulated because, in substantially altering what is "natural," they will also change the current social order—for conservatives the desirable "end of history." Fukuyama attempts to locate what is fundamentally human in a genetically derived essence. Such essentialism goes hand in hand with the kind of technological determinism that Feenberg has been at pains to reject. Indeed, Fukuyama's overreliance on a hardwired human essence leads him to avoid considering the social and cultural contexts that shape human actions. For instance, he wants to regulate or ban psychotropic drugs on the understandable premise that they are able to create a false and all too easy sense of self-worth, without the corresponding labor—in Hegelian terms, recognition without the struggle. He never asks why there is a need for such drugs within contemporary society. Instead he simply claims that Prozac is consumed primarily by depressed women who lack self-esteem. Prozac works to raise their serotonin levels so that they resemble those of alpha males. The fact that any need for a serotonin boost might somehow reside within the structures of information and capital escapes him. As Steven Johnson points out, "Sweatshops, urban poverty and stratospheric CEO salaries all alter serotonin levels as well, arguable more effectively than Prozac does. You could make the case that these drugs aren't natural. But then again neither are stock options."[5]

It would seem then that works such as Fukuyama's which condemn biotechnology because it violates nature and the human essence merely confirm that such broad critiques of biotechnology fall into the same trap as older substantive critiques of technology such as Heidegger's. That these critiques often align themselves with conservative moralism gives weight to Feenberg's claim that any radical politics of technology cannot be sustained at this level. However, in contrast to Feenberg, I want to argue that in some senses it can—not through essentialism but instead through looking at how once culturally grounded practices can be transformed, and what this might mean for a culture, or a society as a whole. I will return to this question.

The second type of critique refuses to reject such technology out of hand. Instead it examines the contexts through which applications of biotechnology might unfold. Such approaches offer a critique of biotechnology, not directed at the technology itself, but rather at the manner through which it might entrench current inequalities and power structures. Thus, biotechnology is problematic when it attempts to commodify forms of life (Icelandic DNA) or rework traditional exchange structures within a market paradigm suited to the West (e.g., biopiracy). This second form of critique looks for ways in which such processes might be resisted, indeed resisted through the appropriation of the very same technologies. A classic example of such an approach is Donna Haraway's "cyborg manifesto," which illustrates, in Feenberg's terms, the ambivalence of the cyborg. Feenberg is closer to this second position. While he hasn't dealt with biotechnology in any detail, his writings have generally endorsed writers such as Donna Haraway on this point and Feenberg has argued many times that technologies are not in themselves problematic, but their functioning within a capitalist framework is the chief problem.

Indeed, the posthumanist focus on technological mediation and hybridity provides, according to Feenberg, a useful critique of essentialism, and thus of the attendant normalizing and marginalizing practices that ensue, such as assumptions about women and nature. There are similarities then, between posthumanist theory and Feenberg's critique of substantive theories of technology that place technology outside the sphere of human involvement. He agrees with the posthumanist rejection of a wholly externalized technology or rationality. It dovetails with his insistence that humans can intervene in the design process of technologies and thus influence their sphere of influence. Thus, Feenberg endorses a cyborg ontology up to a point because under these conditions, technology can never have an unalterable *telos* but will always contain difference and plurality. The cyborg already contains the possibility for an alternative modernity.

Yet Feenberg wants to distance his work from at least some aspects of any cyborg or posthuman critique. Such critiques make no attempt to separate technology from humanity or nature, because all three are part of a mutually constructed network through which we come to know the world. Feenberg avoids taking the notion of technological mediation toward any radical epistemological conclusions, however, because it undermines the capacity for reflexive politics. Thus, as he did earlier with substantive theory, Feenberg takes the critical insights from theorists such as Haraway and Latour while resisting any total embrace of their worldview.

A concern for actual intervention into technological practices prevents Feenberg from wholly endorsing posthuman theory. He is critical of it on two grounds. Firstly, decentering humans as simply elements of a cyborg ontology, or as part of a network involving human and nonhuman actors leads to a kind of performative contradiction, where it becomes difficult to determine the quality, identity, and status of any "act." As Feenberg puts it, "How . . . can we talk about actants without using the language of modernity in which the human and the natural are distinguished a priori?"[6] The second objection to posthuman theory, in particular Latour's Actor Network Theory, stems from a political objection to the hermetic nature of the network where "a strict operationalism [applies] that forbids the introduction of data that is not effective in the strong sense of decisive for the organization of the network."[7] This bias toward existing power, sanctioned through the criteria of operationalism, removes the grounds through which one might choose to alter the nature of the human-technological network. The acknowledgment of plurality, and the destruction of essences, will not necessarily lead to the reorganization of power.

Thus, while Feenberg partially condones the posthuman critique of any separation between the "human" and the "technological," he is aware that the "antiessentialist demand for permanent contestation, for dispersion and difference . . . cannot provide the basis for a positive approach to technological reform."[8] Indeed, his critique of the political assumptions of posthuman theory resemble Hardt and Negri's critique of postmodernism in *Empire*. Both are aware that contemporary power and domination arise not so much from essentialism and normalization, but from the management of networks of hybridity.[9] Feenberg notes that the results of domination such as nuclear weapons and the pollution of the third world

> [a]re not the products of rigid bureaucracies the authority of which is sapped by a new postmodern individualism, but of flexible centres of command that are well adapted to the new technologies they have designed and implemented. The opposition of these centres must also oppose the present trend of technological design and suggest an alternative. For that purpose it is important to retain a strong notion of potentiality with which to challenge existing designs.[10]

It is the values that would inform such "potentiality" that marks the point where I would differ from Feenberg. He locates social and cultural values at the core of his critique of constructivist theories of technology. In relation to such theories he notes that "the frequent rejection of macro-sociological concepts such as class and culture further armours the research against politics by making it almost impossible to introduce the broad society-wide factors which shape technology behind the backs of the actors."[11] Yet it is precisely these social factors which are subject to a contradictory process of change in the move from modernity toward the posthuman. While he locates much of the political struggle to appropriate technologies at the micro level, he doesn't make clear how, at a broader level, such values have been subject to reconstitution within the contemporary situation. One only has to think of how categories such as "class" and "culture" now appear in a new framework enabled through technology. Class acquires new meanings in a context of high-tech production where manual labor is often displaced, where skills and locales are subject to process of mobility and obsolescence. Culture too changes with globalization, where the more concrete frames that created and sustained culture are being replaced by less tangible information networks. In this sense Feenberg's own critique of Latour's "transcendent localism" also marks the limits of his own work. If a posthuman anthropology such as Latour's Actor Network Theory is, as Feenberg rightly points out, unable to theorize the context for politics and ethics, Feenberg doesn't adequately theorize the grounds through which the values he relies on to guide technology—such as democracy, noninstrumentality, and so on—operate in the contemporary context.

In other words, if we are to engage with the question of the posthuman, we may need to ask broader questions than Feenberg has come to allow in his approach to technology. This sends us dangerously toward the kind of overdeterminism and essentialism that he criticizes, but I believe that such questions can be posed in a way that avoids such charges, and furthermore provides a more thorough basis for a reflexive ethics than is allowed through Feenberg's two-leveled process of instrumentalization. While I am aware of the dangers of turning the multiple practices that constitute "biotechnology" into a monolithic structure, it may well be that, analytically, this is necessary in order to examine the cultural framework that biotech practices move us toward.

Biotechnologies propel us toward the posthuman in a particular way. Such technologies are part of a constellation that includes scientific curiosity (as well as hubris), huge amounts of private capital, and age-old desires to transcend our natural limitations. These forces tend to stack the deck against any attempt to open a space for a discussion of the long-term significance of the biotech revolution. Indeed, within this framework, *biotechnology always offers a way out of any particular ethical dilemma we might have.*

As an example we can take the current situation regarding embryonic stem-cell research and practice. Early research in mice reveals a range of possible therapeutic applications (a mouse with spinal injuries was able to move again after eight weeks of stem-cell therapy). All that is needed for therapeutic cloning is some skin and a human egg. Putting any qualms about this process aside for a moment, scientists tell us that there are simply not enough eggs ("spares" from the IVF program) available for research purposes. What to do? The answer would most likely involve the exploitation of poor women, probably in the third world. Such a dilemma touches on many peoples' concern with biotech—the use of some people as resources for the technological enhancement of others. However, biotechnology may have an answer to this dilemma. Writing on this subject in *The Nation*,[12] Ralph Brave reports that Nobel laureate Paul Berg claims that science soon ought to be able to isolate the biochemicals in a egg and transfer them directly to the skin cell, thus creating a new egg from a skin cell. No need to collect eggs for embryos when you can create new ones from skin cells.

However, as soon as biotechnology "solves" one problem another is created. For as soon as you can create embryos from skin cells you move one step closer to removing humans from the reproduction process altogether. If it becomes possible to create embryos from skin, all you need is an artificial womb (not in itself that difficult to create) and you have isolated human reproduction from human embodiment. Thus, for every problem that biotechnologies create, they find a solution. However, as Brave notes it is a solution where "the moral quandary has been replaced by an extracorporeal biochemical process, no longer strictly defined as human."[13] In theory, we can keep solving these problems until we find ourselves in a world we cannot live in. The question of the posthuman rises at this point—how many times can we resolve problems through the technological simulation of embodied processes and still remain "human"?

Such a dilemma invokes the contradictions that arise through technologically reconstitutive practices. Technology can liberate us at the same time as it undermines the conditions that might make such liberation meaningful. Our relation to biotechnologies may need to encompass what I regard as properly ontological questions, and not simply be approached, as Feenberg does, at the level of ontic ambivalence.

This is the posthuman challenge to Feenberg that cannot be answered in terms of the cyborg, the network, or through his critique of them. The question remains, at what point in the biotechnological process might we want to intervene? In some ways, Feenberg's struggle to avoid what he sees as the determinism of thinkers such as Heidegger has meant that he has foreclosed the possibility of thinking about a widespread shift in the frames through which we come to understand the world. For instance, Feenberg is critical of Heidegger's ultimately deterministic attitude toward technology. To avoid this determinism, he distinguishes different social formations (modern and premodern) historically, rather than ontologically, as Heidegger does. Feenberg writes that "I distinguish the premodern from the modern historically, rather than ontologically."[14] While not wanting to endorse entirely Heidegger's ontological periodization as it stands,[15] I would argue that Feenberg undercuts the possibility for a critical reflexive approach by not allowing for an enlarged sense of ontological, as well as historical difference. To say this is not to claim that we move entirely from one kind of ontological frame into another. Rather, it is to claim that at any time social life is composed of the intersection of a number of such frameworks, each more or less abstract in terms of how subjects and objects are constituted. The posthuman represents an emerging framework that stands at the farthest degree of abstraction. Biotechnology promises to be one of the chief means of reconstitution of the way in which we engage with the world and increasingly comport ourselves within this framework. Our sense of embodied self, our relations with others, our relation to death and temporality, may be radically altered if biotechnology allows the transcendence of what were once regarded as natural limits—limits that also ground human culture and society.

Can we speak of a biotechnology as one of the factors driving a cultural shift without lapsing into determinism? Arguably, in an effort to avoid the latter, Feenberg has not been able to explore the former. At what point does the capacity for subversion of the dominant technocratic values become secondary to the possibility of a more fundamental cultural shift? In demonstrating the contingency of technology, Feenberg sets up a dichotomy of freedom against technological constraint. Whereas writers such as Ellul, according to Feenberg, see technology creating an increasingly deterministic future, Feenberg counters by arguing:

> New degrees of freedom open up inside the system at critical junctures and points of passage. Public interventions sometimes succeed in bringing advanced technologies to a halt or changing their direction of development. Fatalism is wrong because technology is equally undermined, crisscrossed by a variety of demands and restrictions to which new elements are constantly added in unpredictable evolutionary sequences.[16]

Up to a certain point one could hardly take issue with this statement. Yet it allows for a certain equivocation in relation to how we might want to make statements now about emerging technoscientific practices and the values that might be called upon to underwrite any future society where such practices occupy an increasingly central place. As Caddick has written:

> How are we to ground (importantly how are we to conditionally ground) our judgement about what it is to be human, and what may be good or bad for us? If

> it is relatively easy to be against human annihilation, as in . . . [technologies such as] Star Wars, what about human cloning, and the need I hear for thousands of body parts, a lack of supply soon to be overcome by the production of transgenic pig organs? . . . [O]n what grounds might some of the practices of the technoscientific age be judged objectionable; to be held abject?[17]

It is not a question of protecting some kind of essential "nature" as Feenberg has remarked when commenting on the work of Latour and Haraway. However, it is the long -held cultural reference points that ground our values and our sense of being-in-the-world which are at stake, rather than nature. Such grounds may be contingent, but they are not easily discarded without a cost. I would argue that Feenberg's work ought to be complemented with a more wide-ranging understanding of the contribution of technology toward a wider social and cultural transformation. Let us look at both this process and Feenberg's work in more detail.

As we have seen, Feenberg attempts to mediate between substantive theories, which ascribe an instrumental *telos* to technology, and constructivist theories, which emphasize the local and contingent role of technology within a network of human actors and cultures, but which lack the enabling conditions for a radical technological politics. The object is to encompass the critical impulse of the former with the potentiality contained within the latter. His two-stage theory of instrumentalization, represents a comprehensive attempt to negotiate between what Feenberg regards as a disempowering universalism and a constructivist particularism that lacks sufficient critical resources. I will explore this theory in a moment, but first I want to look a crucial figure that, for Feenberg, enables a sense of agency to be injected into the technological process—the specific intellectual. Given that much of his work is spent attempting to negotiate the poles of universalism and particularism it is not surprising that he increasingly turns to the specific intellectual, a figure that for Foucault stands at the transition point between these poles.

Feenberg's Technological Politics

Technical networks are not the overdetermined structures imagined by substantive theorists. On the contrary, they are subject to human agency, and for Feenberg, it is those who design and implement technical systems who can play a key role in the reappropriation of technology away from narrowly technocratic or economic goals. He relies on Foucault's concept of the "specific intellectual" to illustrate how strategies by specific actors can redetermine the technical network. Thus:

> [S]pecific intellectuals constitute a new class of heterogeneous engineers whose tactical labours in their technical fields extend the recognized boundaries of networks, often against the will of the managers.[18]

Like Foucault, Feenberg recognizes that intellectual practice has become a positive constituting force. The specific intellectual works "within specific sectors at the precise points where their own conditions of life or work situate them."[19] Feenberg comes to increasingly draw on Foucault's theory when making claims for the power of those involved in technical processes to reappropriate the technology via a process of "subversive rationalization."

However, Feenberg does not ask deeper questions about the conditions of possibility for the specific intellectual. Like Foucault, he seems happy to register their existence empirically, then proceed to consider their strategic possibilities in terms of a technological counterhegemony. Yet the significance of the specific intellectual lies not simply in what strategies they have available to them, but the changed relation of intellectual practices to the social realm. The very conditions that have produced the specific intellectual—the expansion of intellectual practice—have also contributed to a widespread reconstitution of social life. The social structure is now increasingly framed by the practices of intellectual labor, displacing prior frames of work and culture. The predominant form of intellectual labor is instrumental, rather than broadly interpretative. Indeed, "individual" creativity has transcended the bounds of specific institutions to become an increasingly dominant way of seeing the world. That is to say, a world in which everyday social and cultural lives are created precisely, in a creative synthesis of media and information. If the specific intellectual is to play a central role in making technology more democratic, we need to theorize the changed relation of the intellectual practices to the social realm.[20]

Intellectual practice constitutes a specific mode of sociality as well as a distinctive way of taking hold of the world. Both of these are constitutively abstract in their mode of engagement—thus, via a medium such as writing or the Internet social relations are able to take place in the absence of the tangible other. In addition, intellectual practice renders abstract its object of inquiry—cultures become systems of information, natural objects are taken apart and reconstituted, the human body becomes a genetic map, and so on. Traditionally, intellectual practices intersected with more concrete ways of life and provided a source of critical and cultural interpretation for the culture that supported it. The shaman, the priest, the philosopher, the political theorist, and the humanistic scientist have all stood in such roles.

Today, however, the more abstract ways of knowing and engaging with the world enabled by intellectual practice have become more dominant. Instead of intersecting with less abstract modes of engaging with the world we find that the scope and constitutive power of intellectual practice has been radically enhanced via the technosciences and the collapse of cultural/moral frameworks that might have set limits on intellectual activity. In addition, the degree to which intellectual practices have come to the center of daily life has increased. More and more of our lives are constituted through some kind of intellectual practice—culturally through the use of media and information, materially through the replacement of natural environments with technoscientific ones. As John Hinkson observes:

> [T]he information revolution progressively remakes the social in the image of the social exchange typical of intellectual practice. Arguably this transformation is more encompassing than the relation of markets, capital, and labour. Cutting across these capitalist relations is the new contribution of intellectual practice embodied in high-tech. The work of the intellect takes apart the work of the hand in economic reproduction, the body in reproduction, and bodily presence in social relations.[21]

That our taken-for-granted ethical points of reference, based in mutuality and presence, are superseded by more abstract relations means that it is vital to understand the nature and limits of a way of life based around the intellectual practices. In other words what are the implications of the expanded role of the intellectual practices on the social realm? The rise of the "knowledge society" carries its own contradictions in that it has ushered in the structural conditions where knowledge no longer necessarily equates with progress, but equally with instability and obsolescence. Steve Fuller comments on the effects of a rise in the number of intellectually trained. He notes that

> the overall increase in high-skilled labour means that the value of being highly skilled declines, which in effect makes any given member of the "elite" more dispensable than ever. . . . In that respect, informationalism's openness to "lifelong learning" backhandedly acknowledges the inability of even the best schooling to shelter one from the vicissitudes of the new global marketplace. Education, though more necessary than ever, appears much like a vaccine that must be repeatedly taken in stronger doses to ward off more virulent strains of the corresponding disease—in this case, technologically-induced unemployment.[22]

"Lifelong learning" in the contemporary context actually indicates the need for constant retraining in order to ensure a small degree of security in the postindustrial economy. This structural shift, which replaces more concrete forms of labor with those governed by the intellectually trained, indicates a different kind of universal process, which underwrites the specific intellectual. Undoubtedly, the specific intellectual is well placed to strategically intervene in technical processes, but there is a need to ask other questions about the long-term viability of a society largely reconstituted through the practices of specific intellectuals.

This leads to the issue of the different forms of reflexivity that might determine how we are to approach the question of the "'posthuman." One approach, favored by Feenberg, places an undue emphasis upon the autonomous capacity of the social actor to master specific domains, whether it be the use of technique by self-active subject, or the emancipatory potential of the specific expert to intervene and reconstruct specific environments. This approach to reflexivity remains within a single theoretical level in that it fails to consider how these social actions are framed within different constitutive forms.

As an example one might consider the category of the individual as they operate within the different structural paradigms of modernity and postmodernity. One can compare the modern individual existing with a relative degree of autonomy with the individual within postmodernity where a more heightened form of individuation comes into being. In the latter, the subject is able to comprehensively transcend the relatively fixed structures based in place and the concrete other. The intellectual practices and their role in creating a more abstract form of being via the information society and/or the technoscientific reconstitution of nature, allow the individual to make/unmake her- or himself from the realms of media, information, and/or the market. Such realms facilitate the overcoming of the constraints of embodiment, enabling the subject to engage with the world outside of a specific location in time and space. Within this more abstract mode of engagement, the tangible presence of the other is no longer a defining dimension of the causal setting for interaction. The individual has the capacity to "stand over" the social other, rather than be constrained by it, as in modernity. One result of this is that individuals may comport themselves differently within this more open structure. This historically novel situation cuts across how humans have lived together and generated a sense of social and ethical responsibility in the past. The second approach to reflexivity would ask the extent to which we can appropriate and subvert technologies toward democratic and political ends in the light of this *structural shift* toward individuation. It would ask whether it is necessary to preserve, as a cultural and ethical necessity, modes of engagement that are, in theory, now technologically disposable.

It is this structural shift toward more abstract modes of engagement, a shift made possible by the expanded scope of the intellectual practices (expanded in their practice if not necessarily in their interpretative capacity), that Feenberg is both aware of and yet curiously underplays in his "dialectic of technology."

Feenberg's two-stage theory of instrumentalism attempts to dialectically incorporate the primary orientation of technology toward a negative deworlding, with a secondary process whereby technologies are reappropriated within social contexts. Feenberg argues that technological decontextualization serves the interests of capital, as it strips subjects from known contexts into more abstract contexts which are in the first instance able to reify human activity. Part of any radical political project involves a subsequent process of recontextualization. Such a recontextualizing practice is

> orientated toward a wide range of interests that capitalism represents only partially, interests that reflect human and natural potentialities capitalism ignores or suppresses. . . . These interests correspond to the lost contexts from which technology is abstracted . . .[23]

Feenberg's dialectic, especially when stated at this level of generality, is entirely suited to cope with a contemporary politics of technology. It avoids the generalized negativity toward technology of writers such as Marcuse and Heidegger, while it retains the wider political impulse necessary to orient any actions within a discreet technical network. However, I still have some concerns. The first is the degree to which Feenberg's dialectic can deal with a potentially transformative technological practice such as biotechnology. The discrete form of technological intervention he advocates (via specific intellectual practices) makes it difficult, as we saw with the case of embryonic stem cells, to ask, "At what point do we move beyond a desirable form of being 'human'?" The second and related issue is that Feenberg underplays the role of contradiction—how the de-worlding tendency he names as part of the primary instrumentalization process is often embraced as emancipatory, as if the subject can operate unfettered *and unchanged*, free from the older frameworks which constrained it but also contextualized their lifeworld.

The contradictory way that abstract relations are lived *as if* they were "concrete" both grants meaning to the abstract relation and threatens to erode such meaning. As Alison Caddick notes:

> [W]here we have no self-reflexive access to those levels of culture that implicitly anchor our humanity, it is exactly the deeply taken-for-granted desires so strongly anchored here that might catapault us beyond those elements of culture which have historically sustained them.[24]

It is difficult to think of something like the kind of social transformation that biotechnologies shift us toward. It is also difficult to think of this as a contradictory process. It is far easier to point to successful examples or imagined possibilities of individual transcendence over physical or cultural constraint. However, I would argue that we need to think of the long-term consequences of such transformations, and not simply base any discussion of technology, and especially biotechnology, in terms of available political options. If we can return to the example of stem-cell research. There is no doubt that biotechnologies alter the framework though which we value the notion of embodied life. Think of the way processes such as organ transplantation render more abstract our relation with our own and other bodies, or how IVF delivers an end result via the abstraction of the body from the process of reproduction. On an individual basis one could argue that we can easily accommodate such changes, but how does the biotechnological reconstitution of embodied processes affect the cultural meanings we attach to things such as childbirth, human enhancement, or the cultural significance of life itself? Because we have never been challenged at such a fundamental level before, we have never had to consider the assumptions we hold about the significance of embodied life. Most people hold a number of competing assumptions together. On the one hand, we hold the notion of an individual right to obtain a better life if it is possible. On the other, a sense of unease that biotechnologies grant us the possibility of achieving our goals through means that either dismember or abstract our bodies from the process of life, health, or death.

The notion of a shift in the broad assumptions or beliefs a culture holds about something is often difficult to trace. It invites the challenge of essentialism, the very thing that Andrew Feenberg has rejected since his work on technology began. We all implicitly know, however, that changes in cultural understanding do occur. We can think how communications technologies have altered our perceptions of time and space so that we increasingly imagine ourselves as global subjects. Or how the notion of property has changed historically and how it differs from culture to culture—from indigenous relations to land, to feudal relations, to private property, to the abstract concept of intellectual property. These shifts and variables profoundly alter the way we come to understand ourselves and relate to the world. Because biotechnology cuts across many of our assumptions about embodied life and social being it must also change our relation to the world. Yet the process of ontological contradiction means that a practice such as biotechnology has the potential to speak to many of the deep-seated desires for liberation, while undermining them at the same time.

Biotechnology presents a useful challenge to us because more than any other technology it forces us to take our long-held and taken-for-granted assumptions about nature, humanity, and the social and theorize them within a reflexive framework. Much of Feenberg's work has been directed toward such a reflexive approach, however, I believe that he has not fully explored the significance of how the frameworks through which we engage with the world are reconstituted through technology. The capacity for technology to reconstitute human actions and values goes beyond the question of extension, beyond the idea that that technology simply extends the means to satisfy cultural needs. There is an entire tradition of substantive theory that can point to cases where this is clearly not the case. However, Feenberg's focus on *ambivalence*, while drawing attention to the possibility of appropriating technology outside of inhuman, technocratic contexts, remains unable to fully meet the challenge of the "posthuman future" because it conceptually remains within the one-dimensional ontology of that future. While it is true that technologies can be challenged and appropriated through the assertion of equitable social and political values by human agents, the question of how these values are themselves subject to transformation as we move into a more abstract technological society needs more attention. To reflexively engage with such a future means that we have to ask wider questions about the cultural shift enabled through the increasing centrality of technology. Such a shift gives rise to a looser, more abstract mode of being-in-the-world. This means that the settings that have always grounded social life and any sense of a cooperative ethic are destabilized. To what extent we can live cooperatively in the absence of these settings is a question that, at the very least, needs to be considered as we move toward some kind of posthuman future, with the hope that it will not be an inhuman one as well.

Notes

1. Andrew Feenberg, *Questioning Technology* (London: Routledge, 1999), 207.
2. G. Sharp, "Is this the End of History?" *Arena Journal* 19 (2002): 5.
3. For an extended discussion of this concept see Feenberg, 201–25.
4. This term arises from the group of writers associated with the Australian journal of social theory, *Arena Journal*. In particular see G. Sharp, "Constitutive Abstraction and Social Practice," *Arena Journal* 70 (1985): 48–83; P. James, *Nation-Formation* (London: Sage, 1996); and S. Cooper, *Technoculture and Critical Theory: In the Service of the Machine?* (London: Routledge, 2002).
5. S. Johnson, "Goodbye to all that," *Washington Post*, 14 April 2002.
6. Andrew Feenberg, *Transforming Technology: A Critical Theory Revisited* (New York: Oxford, 2002), 30.
7. Ibid., 31.
8. Ibid., 32.
9. M. Hardt and A. Negri, *Empire* (Cambridge: Harvard University Press, 2000), 142.
10. *Questioning Technology*, 32.
11. Ibid., 11.

12. R. Brave, "The Body Shop," *The Nation*, 22 April 2002, 25.
13. Ibid.
14. *Questioning Technology*, 208.
15. See Cooper, chapter 2, for a more comprehensive discussion of Heidegger on these.
16. A. Feenberg, *Alternative Modernity* (Berkeley: University of California Press, 1995), 231.
17. A. Caddick, "I'd Rather Be a Cyborg than a Goddess," *Arena Magazine* 35 (1998): 10.
18. A. Feenberg, "Escaping the Iron Cage, or, Subversive Rationalization and Democratic Theory," in *Democratizing Technology*, ed. R. Schomberg (Tilburg: International Centre for Human and Public Affairs, forthcoming). Also posted at http://www-rohan.sdsu.edu/faculty/feenberg/schoml.htm (accessed 12/1/03).
19. M. Foucault, "Truth and Power," in *The Foucault Reader*, ed. P. Rabinow (Harmondsworth: Penguin, 1984), 69.
20. For an extended discussion of the role of Intellectual Practices in the context of the current university see S. Cooper, J. Hinkson, and G. Sharp, *Scholars and Entrepreneurs: The Universities in Crisis* (Melbourne: Arena, 2002).
21. J. Hinkson, "Perspectives in the Crisis in the University," in *Scholars and Entrepreneurs*, 262.
22. S. Fuller, " 'Making the University Fit for Critical Intellectuals' Recovering from the Ravages of the Postmodern Condition," *British Educational Research Journal* (December 1999): 585.
23. *Transforming Technology*, 184.
24. A. Caddick, "Witnessing the Bio-Tech Revolution," *Arena Journal* 8 (1997): 59–92, 69.

CHAPTER THREE

TRISH GLAZEBROOK

An Ecofeminist Response

I come to Andrew Feenberg's work as a Heideggerian ecofeminist. Heidegger's critique of science and technology has been useful to several ecofeminist writers, some more, some less explicitly. In my case, his work has been more than influential, in fact formative, primarily because of his analysis of the Western intellectual tradition. He argues that this tradition culminates in modernity in a logic of domination. Ecofeminists agree, and add that this logic is phallic. I have furthermore used Heidegger's account of essence to suggest a sense in which essentialism can be thought not biologically but historically as a conceptual resource for ecofeminist alternative environmental epistemologies.[1] I have borrowed ecofeminist strategies of discourse and inclusivity, in response to Michael Zimmerman's worries,[2] to develop a Heideggerian environmentalism that precludes fascism, and I have explored the account of dwelling found in Karen Warren's work in terms of Heidegger's analyses of *Heimat* and *Unheimlichkeit*.[3] Rich and insightful though this work has been for me, I must confess that Feenberg's writing came as a somewhat dramatic revelation to awaken me from my dogmatic slumber, as it were.

My first encounter with Feenberg's writing was in 1995. His "Subversive Rationalization: Technology, Power, and Democracy" shows that technology theorists need not dissociate themselves from praxis as a consequence of the rhetoric that totalizes technology into an ideology of domination.[4] The social and political possibilities for subversive rationalization are an antidote to Luddite passivity and pessimism. Heidegger argues that technology is not just a collection of equipment, but a way of opening reality, of structuring intelligibility, of organizing a world. Feenberg argues further that technology is a source of public power. His analyses of the technology-based social infrastructures in which users are embedded are more helpful than Heidegger's "*Technik*," if one seeks not just diagnosis, but practical possibilities for resistance and democratization.

We are on one hand cornered by technology—as Hwa Yol Jung put it recently, "Technology is somehow being forced on me, and there seems no escape."[5] It is not just that he feels ongoing pressure to use e-mail. Rather, technology seems inescapable as a system, despite the capacity to resist specific technologies. Erika, a high school student with whom I play djembe, complained during a break we took between rhythms: "I hate computers. But I finally got a laptop." Why? "Peer pressure. All my friends did, and I want to be in classes with them." At her school, students with laptops are in separate classes from everyone else. They take notes on their laptops, do projects, and connect to the Internet during class. Is not to have the technology to put oneself at a disadvantage? The technology segregates—apparently democratic in that all students are free to bring a laptop, it actually privileges, and reproduces race, wealth, and class lines in doing so. But Erika doesn't seem to mind that so much. She hates computers "because they are supposed to be simple, but they make things more complicated. There's crashes, and losses of work, and things never go like they should—a ten-minute task seems to always take half an hour—and I have to lug it around all the time." I understand—I could easily occupy my whole day with "labor-saving" technologies. My labor is not "saved" at all, but diverted machine-ward.

Furthermore, consider the bicycle. Feenberg shows how factors external to the design process influenced the technical development of the bicycle away from racing bikes toward the "safety" bike ridden today.[6] A more worrisome history of the bicycle is its metamorphosis into a recreation device, and its virtual elimination from transportation in North America. Like the roads themselves, which used to serve a variety of social and practical functions, the human desire to get about has been almost entirely taken over by the car. *Star Trek*'s Borg may be right: "Resistance is futile. You will be assimilated." I lived in a small town in the United States for six years without a car, and very few people seemed to understand the choice. Some would say, "But of course you can bike to class, because you live so close." I would say, "No, I live close *in order that* I may bike to class." Others kept offering me their cars, as if some extraneous factor prevented me from having a car, which I naturally would have, if I could. When I wanted a van to drive a group of students on a research trip, the Director of Humanities asked, "But who will drive?" He assumed that not owning a car indicated no license. What alternatives are there to the car? "Mass" transit is only available where there are masses, that is, in larger urban areas, and car manufacturers promote a conception of the automobile in which mass transport belongs to the underclass by deprivation and necessity, not by choice. I eventually got a car. At a friend's house, we both complain about having our cars in the shop. I say, "Sometimes I think all we do is work so we can afford to keep our inadequate and second-hand technology at some minimal level of repair." She accuses me of cynicism. I've been reading too much Heidegger. Could more democratic access to technology, creative strategies of resistance or intervention, or subversive rationality, change things? I am, like Hwa Yol and Erika, caught in a web.

On the other hand, I find myself embedded in technology in liberating ways. For example, though not recommended as the choice for everyone, I am of a certain not young age, yet childless. I have made this choice at a deeply personal level, not so much as a career strategy, but as an ecofeminist conviction. That is not to say that feminists and mothers are exclusive or oppositional groups, but that I (against Feenberg's more recent arguments that overpopulation is not inherently environmentally destructive)[7] believe that there are just too many of us on this planet, and that I do not wish personally to enlarge the population that is most disproportionately (ab)usive of resources and (counter)productive of waste. A phenomenology of the lifeworld quickly uncovers the ways in which technology has increased rather than confined my options. Some feminists have warned, however, that new reproductive technologies may offer some women better or at least more options, but that an increased set of options between which one can select does not necessarily entail the empowerment brought by authentic choice, nor guarantee democratic access to such choices for all women.[8] Coke or Pepsi? These are options, not real choices.[9] Options push into the background or block access to real, authentic choices, such as the choice to reclaim my psychic life from its colonization ubiquitously and insidiously by corporate logos.[10]

When it comes to technology, Heidegger thinks that there is a difference between "catching sight of what comes to presence in technology . . . [and] merely staring at the technological,"[11] "thinking and mere wanting to know,"[12] ongoing activity and mere busyness,[13] and "genuine knowledge . . . [versus] increasingly comprehensive and secure mastery of objects."[14] How is this contrast to be understood praxically? Technology, as a way of revealing being (that is, a way of thinking, rather than a collection of equipment) is aligned with the latter disjuncts by Heidegger, and is allegedly bad, while the first of each pair of disjuncts is supposedly good. This clear division is awkward. One cannot make a choice between disjuncts in the Heideggerian either/or of technology, for one's cultural and historical location embeds one inextricably in technology. *Especially if one wants to engage in a cultural critique of technology.* That task requires word processors, books, a host of other technological contraptions, and more importantly, a background commitment to the technological *Ge-stell* about which Heidegger warns. One must precisely be oblivious to the metaphysical and ethical (that is, pertaining to *ethos*) questions concerning technology while engaging in such critique. Were one to model one's life after the simple peasant woman of "The Origin of the Work of Art," who "takes off her shoes late in the evening, in deep but healthy fatigue, and reaches out for them again in the still dim dawn,"[15] or "the farmer, slow of step,"[16] who draws furrows through the field at the end of "Letter on Humanism," one should hardly be able to contribute to the philosophy industry's critique of technology.

But this is not the choice at stake in Heidegger's analysis. Rather, the disjunction is between unthinking complacency, and thoughtful critique. As a Heideggerian ecofeminist, then, I ask Feenberg, What are the social and praxical functions of a critique of technology? Are democratic interventions chronicled as models for other resistance? Is there something we need to know about technologies and technocracy in order to democratize them? This is the question of the relation between philosophical analysis and political resistance—the politics of knowledge. It is precisely here, at the connection of reflection and practice that ecofeminism has made its home, a home characterized by *Unheimlichkeit*, that is, a displacement through unanswered questions that makes thinking possible.

There is a further worry here. Ontic choices (that is, choices about beings, e.g., coffee or tea?) made in the quotidian do not necessarily follow from some ontological and a priori unfolding of being. Rather, it may well be through such ontic choices that a destiny of being is played out. In less Heideggerian terms, ideology does not inform practice a priori; rather, ideology itself arises from the lifeworld. Heidegger offers an either/or in which one is either blindly driven by technology as progress, or pessimistically helpless once one has seen through to the threat of technology. Feenberg thinks that this is because in Heidegger's analysis, "modern technology is seen from above," and Heidegger thus does not uncover "its disclosive significance for ordinary actors."[17] The poverty of this either/or is particularly marked when one turns from abstract, theoretical reflection on "the essence of technology" to praxical questions concerning concrete and particular manifestations of technological thinking, such as new reproductive technologies, or corporate practices in consumer culture. An analysis of the abstract threat of the essence of technology simply does not help. Ecofeminism offers a release to some extent from this Heideggerian bind, because ecofeminism offers a philosophical and theoretical conceptual framework consistent with Heidegger's analysis of technology, which I find persuasive and convincing, yet coupled with social and political strategies to resist ideologies and practices of domination.

Ecofeminism has, however, been charged with irrationalism,[18] because its alternative epistemologies resist a totalizing homogenization of rationality into scientific objectivity. Feenberg's talk of subversive rationality is the promise of a response to this charge. For Heidegger and ecofeminists argue for a rationality alternative to its modern realization in science and technology, but it is Feenberg who provides ways to think about what that might mean, what it might look like, and how it could be praxically effected. Feenberg shows that subversive rationalization is not irrational. The Heideggerian contrast between "catching sight of what comes to presence in technology . . . [and] merely staring at the technological" is the difference between unthinking immersion in technology, and self-conscious thoughtful reflection on the technological context in which one finds oneself inextricably embedded. Feenberg gives a sense of what it looks like to be thoughtful in this way, though I still pose to him the question of the politics of knowledge articulated above. In fact, I now pose the same question from the other direction: What is the relation between the rationalities at work in technological systems, their power as ways of thinking, and technological praxes?

Accordingly, I read Feenberg with high expectations, and I apologize to him if the order seems unrealistically tall. My further reading of his work has not yet, however, disappointed me. In this chapter, I will therefore push his thinking in a spirit of critical celebration by responding to his work as a Heideggerian ecofeminist, with several goals in mind. First, the quibbling scholar in me cannot resist this opportunity briefly to set the record straight somewhat with respect to Heidegger. Feenberg's criticisms of Heidegger are singularly deep in their articulation of problems commonly diagnosed as belonging to Heidegger's analysis of technology, that is, his essentialism, determinism, and pessimism, yet Feenberg is perhaps more Heideggerian than he admits. Secondly, I want to point out some fundamental moments in Feenberg's thinking that are shared with feminism. Finally, I intend to give some rejoinders to aspects of Feenberg's work from an ecofeminist perspective. I will layer these three objectives through thematic treatment of four specific topics: anti-essentialism, Feenberg's AIDS example, the body as a political site, and the relation between ideology and praxis.

Anti-Essentialism: Nurture over Nature

One of the things Feenberg has most obviously in common with feminism is his rejection of essentialism. Simone de Beauvoir began the so-called second wave of feminism with her claim in *The Second Sex* that "[o]ne is not born, but rather becomes, a woman."[19] That simple observation ushered in the sex/gender, a.k.a. nature/nurture, distinction. Its purpose was to disrupt the biological determinism that justifies women's oppression by making it appear a destiny. If gender is socially constructed, then women need not be reduced to their reproductive capacities, as if motherhood were their biologically and/or divinely guided fate. Gender constructed by social praxes is malleable. Oppression is the consequence of social and political relations of power that could be arranged differently. Accordingly, the feminist argument for the social construction of gender was not just a correction of mistaken ontology, but rather a political strategy to make change possible by suggesting that gender need not be played out in the ways it has been. In Feenberg's terms, the second wave of feminism was a subversive rationality of gender.

Likewise, Feenberg resists technological determinism and essentialism. He has long been sympathetic to constructivist accounts of technology, because he has long insisted that technology is a dimension of culture with implications for the distribution of political power. In *Alternative Modernity*, he argues "from the constructivist premise to the possibility of reshaping the technical world around us,"[20] and in *Questioning Technology*, constructivism is for him an antidote to technological determinism. Constructivism argues that choices between competing designs depend not on technical factors or economic efficiency, but on fit between the devices and the interests of the groups that influence design. What singles an artifact out "is its relationship to the social environment, not some intrinsic property."[21] Feenberg's intention is to build from the constructivist position a hermeneutics of technology. He is interested not just in technically explicable function, but in hermeneutically interpretable meaning.

Like second wave feminists, Feenberg is not simply out to use constructivism to correct an error with respect to ontology. His *Alternative Modernity* intends to establish three points: that "technological design is socially relative; . . . the unequal distribution of social influence over technological design contributes to social injustice; and . . . there are at least some instances in which public involvement in the design of devices and systems has made a difference."[22] Together, these three points articulate a resistance strategy to promote social change, from injustice toward a more democratic world. What does "a more democratic world" mean here? As resisting essentialism meant for feminism a transition from woman as subjected by her body to woman as embodied subject, so Feenberg argues for the possibility of an alternative modernity in which human beings participate in technology as rational agents, that is, self-determining subjects, rather than simply being subjected to technical codes and practices controlled by an elite group.

AIDS: *Subjects Without Subjection*

For example, Feenberg shows how AIDS activists refused to accept their passive role as patient in the medical-industrial complex, and instead actualized their participant interest. They intervened in the system as subjects. If the sick are poorly informed, then using them to test experimental drugs violates their rights. If, however, they are well informed, it is paternalistic and disempowering not to include them in decision-making processes concerning experimental drugs. In *Alternative Modernity*, Feenberg argues that the "ethical obligation of medicine is best fulfilled not by prohibitions, but by ensuring that patients are well equipped"[23] to make judgments about their own health care. Later, in *Questioning Technology*, AIDS sufferers did more than change their role with respect to medicine in Feenberg's analysis: their "struggle represents a counter-tendency to the technocratic organization of medicine, an attempt to recover its symbolic dimension and caring functions through democratic intervention."[24] As public users did in France with the Minitel, argues Feenberg, AIDS patients intervened in a technocratic system in order to accommodate that system to excluded interests. Accordingly, they made the medical-industrial complex more democratic—they reinvested medicine with the value of patient care, and thus returned it, to some extent, to its function of serving the people.

From an ecofeminist perspective, however, questions remain about this democratic intervention into the medical industry. According to Feenberg, the original design of the system "reflected the interests and concerns of the technical and administrative elites,"[25] and the users resisted by imposing another layer of function on the medical system that reflected their interests, which were excluded by the original design. The success of this intervention was possible because "AIDS patients were 'networked' politically by the gay rights movement even before they were caught up in a network of contagion."[26] Here's the worry: insofar as the gay rights movement itself brings about social progress, it may not be irrelevant to its ability to do so that its proponents, particularly its organizers, are for the most part white males, already situated in the technical code in empowered ways. The benefits of their intervention accrue to a larger, more diverse group, but at issue here is the capacity to intervene.

Think, for example, of a specific technology that suddenly received widespread acceptance through the gay rights/AIDS movement: the condom. That many (especially young, without the heterosexist, patriarchal protection of marriage) women's lives have been hampered, damaged, or destroyed socially, economically, and experientially, sometimes to the point of death, by unwanted pregnancy, brought about no positive, progressive use of the condom. AIDS activists were powerful enough to make condoms not only okay, but also indicative of social competence, that is, "cool," even against such powerful social influences as the Catholic Church. Is the AIDS example of democratic intervention, then, the story of good (democracy) winning out over evil (an oppressive technical code), or the story of a group *already empowered within and by* a system correcting an oversight of that system? Once they voiced their exclusion, it seems to have been immediately clear to the "technical and administrative elites" that here was a group whose rights had been overlooked. Is this example a straightforward success story, or confirmation that the technical code accommodates the changing needs of an elite?

Nonetheless, Feenberg shares with feminists a repudiation of essentialism as a kind of reclaiming of human subjectivity, from patriarchy and technocracy, respectively. Furthermore, he suggests that one way to intervene successfully in the technical code is to redefine, as public, interests that have traditionally been taken as private. For example, the movement by the disabled for barrier-free design was a call for a simple design change, the sidewalk ramp. In the existing technical code "disabilities were regarded as private problems."[27] Demand for mainstream social participation by the disabled population made their needs appear rather as matters for public policy. Likewise, feminism has made advances by blurring or breaking down the distinction between public and private. For example, insofar as marital rape is recognized to exist, this is because feminist voices demanded that this "private" concern be acknowledged and remedied against in law. Conversely, privatization of environmental responsibility smokescreened corporate accountability when a corporate program, involving hundreds of millions of dollars of advertising space, diverted "environmental pressures away from business and toward individual action."[28] Corporate intervention defined environmentalism as a private responsibility, but democratic intervention entails redefining issues as matters of public concern.

Hence, both Feenberg and feminism are in at least this way at a distance from American traditions of liberalism. It may be that putting more rather than less of private life into public view will be liberating with respect to both gender and technocracy. In this sense, subjects who redesign technical codes democratically cannot be the isolated solipsists of Cartesian subjectivity. Like feminist subjectivities, Feenberg's participants in the technical code are embodied selves, already in a world that is a public space they share with others.

Political Sites: Nature and Nurture

Feminists are not, however, entirely satisfied with the view that gender is a social construct. Girls would not be nurtured into the gender roles of femininity, resistible more or less but not in toto, if there were not the tell-tale marks of femininity on their body. Social construction is underpinned, however vaguely, by biology. This is not to say that biological determination of sex is always simple, but rather, that social constructions start somewhere. At the same time, there is no access to biology unmediated by social construction. Menstruation, for example, is a biological fact. This does not mean that all women menstruate, but rather that menstruation is a bodily function. When women menstruate, the experience is structured by cultural context. One cannot narrate any woman's experience of menstruation by giving a strictly physiological account, or by leaving that account out entirely. Useful, then, though the nature/nurture distinction has been, feminism has moved *beyond* it. Biology and social construction are inseparable, and hence a new view is emerging in feminist thinking: woman's body is a political site. This thesis can be articulated in at least three ways, though if I am right, more should emerge.

First, women's bodies are the site wherein the politics of rape, pornography, and prostitution are played out. The issues at stake here concern in part access and ownership. For example, marital rape is a prosecutable offence when a woman, not her husband, is considered to own her body and sexuality. Secondly, as Cixous argues, woman enters history through writing.[29] The self-articulated and autonomous identity that emerges in women's writing includes her definition, delimitation, and understanding of her body, for example, what her sexuality means to her. Hence, thirdly, woman's body is a political site because feminist philosophy is a philosophy of embodiment. It denies a Cartesian mind/body separation, and through this denial, throws over a separation of the theoretical from the practical. Feminist issues are not just ideological, but practical. They concern to what use our bodies are put, by whom, and for whom. In particular, woman's body as mother, as pregnant, laboring, and lactating, is at issue in feminist ethics of care. This is not because women are necessarily mothers, nor does it imply that women who are not mothers are deficient. Rather, woman's potential for giving birth (en)genders ethical practices in the ways Carol Gilligan, Nel Noddings, and Joan Tronto have made clear.[30] Maternal practices are an alternative ethic to the (phallic) liberal tradition of rights, equality, and justice.

Similarly, Feenberg appears to have become over time dissatisfied with the either/or of essentialism/constructivism, and to have opted instead for a both/and in which he draws from both sides. This is a significant development in *Questioning Technology*, wherein he identifies his analysis of functional constitution with Heidegger's notion of enframing, and says that such constitution as function consists in "four reifying moments of technical practice."[31] Previously, one could not imagine Feenberg talking in such apparently essentialist ways. Yet this analysis does not reduce technology to function, because to do so would be to strip it of value and social context. Accordingly, Feenberg presents a two-level theory of technology. The first level is functional constitution, the aspect of the essence of technology that is its "primary instrumentalization,"[32] which echoes the essentialist analysis in terms of function and the value of efficiency.

The second aspect is "secondary instrumentalization,"[33] the way in which technologies are realized in a functional context wherein technique is integrated with the natural, technical, and social environments.

The notion of essence that results from this combination of thinking through technology in terms of both function and context has much in common with what "*Wesen*" came to mean for Heidegger. As William Lovitt explains, "*Wesen* does not simply mean what something is, but . . . the way in which something pursues its course, the way in which it remains through time what it is."[34] For both Feenberg and Heidegger, "essence" is about location and situatedness. It does not fix what something is in an economy of the eternal, but nor is a thing open to any and all interpretation. Hermeneutic flexibility is bounded by cultural and historical location. Heidegger developed this historical notion of essence elsewhere, but it is precisely in his analysis of technology that he puts it to work. For Feenberg, it is a historical notion of essence that provides the starting point for his two-level theory of technology, and that disrupts technological essentialism and determinism.

Feminists argue that gender is neither just nature nor just nurture, but rather that woman's body is a political site, and likewise Feenberg argues that "technology is neither purely natural nor purely social,"[35] but rather, "a site of social struggle."[36] In particular, it is a site of struggle over the body of the worker. The technical code "privileges deskilling as a fundamental strategy of mechanization from Arkwright down to the present."[37] The technical code entails a Foucauldian discipline of the body to turn the worker into an effective tool. But, argues Feenberg, "work can be redesigned to take advantage of human intelligence and skill,"[38] in resistance to the totalitarian and authoritarian disciplines written historically into the technical code. Instead of reducing workers to interchangeable, docile bodies, technocracy could treat workers as full human subjects in Feenberg's vision. Thus, workers could claim back not just their bodies, but also their minds.

Ideology and Praxis

That is to say, technological systems exert control not just over the body, but also over the mind. A telling example of this is advertising. Advertising is a technocratic mechanism of social control. It is a propaganda machine that seeks to determine tastes, desires, and ideology. This is of course, one of Jerry Mander's arguments for the elimination of television.[39] That is to say, advertising sells not just specific products, but also consumerism. As a mechanism for supporting and directing late capitalist, consumer culture, commercial advertising is, as Kalle Lasn has pointed out, a colonization of the individual's psychic space by corporations. Lasn advocates a kind of resistance he calls "culture jamming,"[40] which is similar to what Feenberg calls "democratic intervention." Both advocate a refusal to be complacent, and therefore complicit. Feenberg opens a conceptual space for resistance to the technical code, "subversive rationalization." The ideological components of praxes are at stake in his analyses. This insight was of course Heidegger's: technology is in essence not some piece of equipment, but a mentality, a way of rendering intelligible, of thinking. Feenberg thinks that there are alternative, more democratic ways of thinking that can underwrite technology.

Likewise, ecofeminists argue that Western history can be analyzed ideologically, and their resulting insight is, much like Heidegger's, that the history of the West has been underwritten by a logic of domination. Francoise d' Eaubonne pointed this out in 1974 when she coined the word *l'eco-féminisme* to argue that the oppression of women and the exploitation of nature are related causally.[41] Rosemary Radford Ruether and Susan Griffin point to Greek philosophy, particularly its dualisms, as a decisive factor in the oppression of woman and nature.[42] Carolyn Merchant and Vandana Shiva point rather to modern science.[43] For all these ecofeminists, what is at issue is a *logic* of domination. Karen Warren defines a logic of domination as "a value-hierarchical way of thinking which explains, justifies, and maintains the subordination of an 'inferior' group by a 'superior' group on the grounds of the (alleged) inferiority or superiority of the respective group."[44] The dualisms of man/nature and man/woman function in this way to justify the oppression of both women and nature in the phallic order, suggests Warren. Elsewhere, she argues for another way of thinking, a peace politics that opposes the -isms of domination. She describes this other way of thinking as a pluralistic theory in process, characterized by inclusiveness, that exposes and challenges practices of power over others, and that is a guide to action which has a central place for care, reciprocity, friendship, and love.[45] Her philosophical practice is grounded in an alternative logic to domination.

Feenberg, however, takes exception, for example, to Heidegger's claim that "[a]griculture is now the mechanized food industry, in essence the same as the manufacturing of corpses in gas chambers and extermination camps, the same as the blockade and starvation of nations, the same as the production of hydrogen bombs."[46] "All," says Feenberg, are for Heidegger "merely different expressions of the identical enframing which we are called to transcend through the recovery of a deeper relation to being."[47] Since Heidegger rejects regression, and leaves no room for a better technological future, Feenberg suggests that it is difficult to see in what such a changed relation to being would consist, "beyond a mere change of attitude."[48] But that is of course exactly the point. As Carol Christ puts it, "We have lost the sense that this Earth is our true home."[49] Contemporary social and environmental crises are not just social, political, economic, and technological, but also spiritual. A change of attitude and ideology *is exactly* what is called for. The change of attitude, for example, advocated in Warren's peace politics is not "mere" at all. What unites agricultural practices with concentration camps is precisely a question of attitude. Ecofeminist thinkers have likened battery farms to concentration camps. Animals are reduced to mere resources and controlled, manipulated, and dominated, in fact tortured, to produce maximum profit at minimal cost.[50] Feenberg's call for subversive rationalization shows that he does in fact see that resistance and analysis take place not just at the concrete level of praxis, but also at the theoretical level of ideology. For the struggle toward democratic access, benefit, and control within the technical code is about how human beings understand nature, other people, and animals. It is about attitudes, that is, ideologies embedded in praxes.

Yet both Feenberg and ecofeminists see, however, that in the end it is practices that must change. So the question remains, will changes in practice follow upon ideological shifts? or, does more effective change begin with practices, and change ideology consequentially? Here, an assumption fundamental to feminism comes into play: theory and practice are inseparably and inextricably embedded. And here, Feenberg's position appears paradoxical. Perhaps the most significant contribution he makes toward an alternative modernity is the theoretical analyses laid out in his texts. Feenberg avoids excessive abstraction, he can distinguish between technologies conceptually, and he grounds his discussions in concrete analyses, but the role of theory *as a resistance strategy* has yet to be worked out. Based on his work, it seems one does in fact have to work both sides of the tracks—technology critics must be broad of vision and narrow of purpose. Feenberg's insights are useful precisely because his visions of subversive rationality and alternative modernity are broadly ideological, while his analyses of technical codes, systems, and practices are narrowly concrete.

Breadth of vision is not exclusively, but is nonetheless also, a resistance strategy. For example, ecofeminists argue that the -isms of domination are bound together in the phallic order. This insight cannot emerge simply *in medias res* with respect to concrete case studies. Sexism, racism, classism, and speciesism are mutually supporting oppressions in a global logic of domination. This is not a totalitarian reduction of rationality to one of its historical manifestations, but the recognition that "development" correlates with "technological development" globally, and that this conflation conduces a global realization of technology in ways that reinforce oppression over democracy. Alternatives to the technical codes of Western capitalism are irreparably eroded, colonized, and displaced. How is resistance possible? First, one must see connections. For example, how are both the Gulf War and rain forest preservation linked to car culture? Once one has stepped back to see the intricate global Web of interconnections (without losing sight of the concrete), resistance is possible. What one sees from the bigger picture is the role of corporate culture in implementing and sustaining the -isms of domination. What is Feenberg's view of the role of corporate culture in technocracy, and its place in his critique?

Feminists and Feenberg agree that resistance is built into the margins. This is the basis of feminist standpoint theory: those oppressed in social hierarchies can see alternatives that may not be evident to those privileged in hegemony, and hence the marginalized are empowered for critical resistance by their very location.[51] Likewise, for Feenberg, a system will always produce the spaces of resistance, that is, subjugated knowledges in Foucault's sense, that, in Feenberg's words, "express the point of view of the dominated . . . [who] are 'situated' in a subordinate position in the technical hierarchy."[52] Hegemony produces its counterhegemony, which brings with it space to resist: "[A] certain margin of maneuver belongs to subordinated positions in the capitalist technical hierarchy."[53] Technological systems have built into them a space for tactical resistance, an "anti-program . . . [that is] not merely a source of disorder but can recodify the network around new programs that realize unsuspected potentialities."[54] The symmetry of program and antiprogram "is the basis of a democratic politics of technological rationalization"[55] insofar as the technical code is "a syntax which is subject to unintended usages that may subvert the framework."[56] These moments are what Feenberg calls "creative appropriations."[57] They are initiated by users who take advantage of the "ambiguity" of technology, its "interpretive flexibility,"[58] by imposing upon it their own conception of its function. For example, women's need for reproductive choice can be piggybacked discreetly on contemporary praxes of "safe" sex. Now it's safe for her too with respect to other ways she may be endangered!

Feenberg suggests other micropolitical strategies and tactics of resistance that are available in the antiprogram. Technical controversies are sometimes created by those affected by technologies in order to influence public opinion in their favor. For example, logging activists' stage nonviolent media events to create mainstream awareness of and resistance to clear-cutting. Innovative dialogues have sometimes been organized by experts professionally involved in the technical code. There may be a gap between the interests that direct a technology, and the specialists who realize it in actuality. For example, scientists' resistance to the LD 50 test, a.k.a. the Draize test, was a central factor in its losing the support of the OECD. These examples are mine, not Feenberg's, but they show that his research program is progressive in Lakatos's sense. It "predicts some novel, hitherto unexpected fact," and "leads us to the actual discovery of some new fact."[59] The new insight of Feenberg's research program is that *oppressive technical codes produce and empower their own resistance*, and this insight can be used constructively to make sense of other oppressions within the technical code, and therefore to empower the subjected toward subjectivity. Feenberg is a promising answer to the ecofeminist hope for a "general reconstruction of modernity in which technology gathers a world to itself rather than reducing its natural, human and social environment to mere resources."[60] Oppressive technocracy itself makes change toward more democratic technocracy possible—

> Where danger is, grows
> The saving power also.[61]

Notes

1. Trish Glazebrook, "Gynocentric Eco-Logics," in *Epistemology and Environmental Philosophy*, ed. Christopher Preston (Albany: State University of New York Press, forthcoming).
2. Michael Zimmerman, "Rethinking the Heidegger-Deep Ecology Relationship," *Environmental Ethics* 15, no. 3 (1993): 195–224, 205.
3. Trish Glazebrook, "Heidegger and Ecofeminism," in *Re-Reading the Canon: Feminist Interpretations of Heidegger*, ed. Nancy Holland and Patricia Huntington (University Park: The Pennsylvania State University Press, 2001).
4. Andrew Feenberg, "Subversive Rationalization: Technology, Power, and Democracy," *Inquiry* 35, no. 3–4 (1992): 301–22, reprinted in *Technology and the Politics of Knowledge*, ed. Andrew Feenberg and Alastair Hannay (Bloomington: Indiana University Press, 1995).
5. Private communication.
6. Andrew Feenberg, *Questioning Technology* (London: Routledge, 1999), 79–80, 88–89.
7. Ibid., 54–59.
8. See especially Robyn Rowland, "Motherhood, Patriarchal Power, Alienation, and the Issue of 'Choice' in Sex Preselection," and Helen B. Holmes and Betty B. Hoskins, "Prenatal and Preconception Sex Choice Technologies: A Path to Femicide?" in Gena Corea et al., *Man-Made Women* (London: Hutchinson, 1985).
9. I owe this formulation of this distinction to Sue Campbell.
10. This is the plural of "logo," though an analysis of corporate semiotics as a *logos* in the pseudo-Greek, Heideggerian sense could be fruitful in another paper.
11. Martin Heidegger, "Die Frage nach der Technik" in *Vorträge und Aufsätze* (Stuttgart: Verlag Günther Neske, 1954), 36, and followed by citation to the English translation, "The Question Concerning Technology," in *The Question Concerning Technology and Other Essays* (New York: Harper and Row, 1977), hereafter QCT, 32.

12. Martin Heidegger, "Die Kehre," in *Bremer und Freiburger Vorträge*, *Gesamtausgabe*, Band 79 (Frankfurt am Main: Vittorio Klostermann, 1994), 71–72, available in English as "The Turning," QCT, 42.
13. Martin Heidegger, "Die Zeit des Weltbildes," in *Holzwege*, *Gesamtausgabe*, Band 5 (Frankfurt am Main: Vittorio Klostermann, 1977), 97, available in English as "The Age of the World Picture," QCT, 138, Appendix 2.
14. Martin Heidegger, *Nietzsche I*, *Gesamtausgabe*, Band 6.1 (Frankfurt am Main: Vittorio Klostermann, 1961), 469, available in English as *Nietzsche*, Vol. 3: *The Will to Power as Knowledge and as Metaphysics*, tr. Joan Stambaugh, David Farrell Krell, and Frank Capuzzi (San Francisco: Harper and Row, 1987), 42.
15. Heidegger, QCT, 34.
16. Ibid., 265.
17. Feenberg, *Questioning Technology*, 197.
18. See for example, Janet Biehl, *Rethinking Ecofeminist Politics* (Boston: South End Press, 1991), 2–4.
19. Simone de Beauvoir, *The Second Sex*, tr. H. M. Parshley (New York: Vintage Books, 1989), 266.
20. Andrew Feenberg, *Alternative Modernity: The Technical Turn in Philosophy and Social Theory* (Berkeley: University of California Press, 1995), ix.
21. Feenberg, *Questioning Technology*, 79.
22. Feenberg, *Alternative Modernity*, 3.
23. Ibid., 116.
24. Feenberg, *Questioning Technology*, 127.
25. Ibid., 219.
26. Ibid., 192.
27. Ibid., 141.
28. Ibid., 61.
29. Hélène Cixous, "La rire de la méduse," *L'Arc* (1975): 39–54.
30. Carol Gilligan, *In A Different Voice: Psychological Theory and Women's Development* (Cambridge: Harvard University Press, 1982); Nel Noddings, "Ethics from the Standpoint of Women," in *Theoretical Perspectives on Sexual Difference*, ed. Deborah Rhode (New Haven: Yale University Press, 1990); Joan Tronto, "Women and Caring: What Can Feminists Learn about Morality from Caring?" in *Women and Values: Readings in Recent Feminist Philosophy*, ed. Marilyn Pearsall (Belmont, CA: Wadsworth, 1999).
31. Feenberg, *Questioning Technology*, 203.
32. Ibid.
33. Ibid., 205.
34. Heidegger, QCT, 3–4, note 1.
35. Feenberg, *Questioning Technology*, 165.
36. Ibid., 83.
37. Ibid., 179.
38. Ibid., 219.
39. Jerry Mander, *Four Arguments for the Elimination of Television* (New York: William Morrow/Quill, 1977).
40. Kalle Lasn, *Culture Jam* (New York: HarperCollins, 1999). Lasn is founder of *Adbusters* magazine, and The Media Foundation, Vancouver, Canada.
41. Francoise d'Eaubonne, *Le Féminisme ou La Mort* (Paris: Pierre Horay, 1974).
42. Rosemary Radford Ruether, *New Woman/New Earth: Sexist Ideologies and Human Liberation* (New York: The Seabury Press, 1975), and Susan Griffin,

Woman and Nature; The Roaring Inside Her (San Francisco: Harper and Row, 1978).

43. Carolyn Merchant, *The Death of Nature: Women, Ecology, and the Scientific Revolution* (San Francisco: Harper and Row, 1980), and Vandana Shiva, *Staying Alive: Women, Ecology, and Development* (London: Zed Books, 1988).
44. Karen Warren, " Feminism and Ecology: Making Connections," *Environmental Ethics* 9, no. 1 (1987): 3–20, 6.
45. Karen Warren, "Toward an Ecofeminist Peace Politics," in *Ecological Feminism*, ed. Karen Warren (New York: Routledge, 1994).
46. Feenberg, *Questioning Technology*, 187. Quoted from Tom Rockmore, *On Heidegger's Nazism and Philosophy* (Berkeley: University of California Press, 1992), 241.
47. Feenberg, *Questioning Technology*, 187.
48. Ibid.
49. Carol Christ, "Rethinking Theology and Nature," in *Reweaving the World: The Emergence of Ecofeminism*, ed. Irene Diamond and Gloria Feman Orenstein (San Francisco: Sierra Club Books, 1990), 58.
50. See Ruth Harrison, *Animal Machines* (London: Stuart, 1964); Cathy B. Glenn, "ReLanguaging Speciesism: Extending Cheney's Postmodern Environmental Ethics," read at The International Association for Environmental Philosophy, 5th Annual meeting, Goucher College, Baltimore, October 7, 2001.
51. See Sandra Harding, "Rethinking Standpoint Epistemology: What Is 'Strong Objectivity?'" in *Feminist Epistemologies*, ed. Linda Alcoff and Elizabeth Potter (New York: Routledge, 1993); bell hooks, "Black Women: Shaping Feminist Theory," in *Feminist Theory: From Margin to Center* (Boston: South End Press, 1984), and "Choosing the Margin as a Space of Radical Openness," in *Yearning: Race, Gender, and Cultural Politics* (Boston: South End Press, 1990).
52. Feenberg, *Questioning Technology*, 111.
53. Ibid., 207.
54. Ibid., 117.
55. Ibid., 119.
56. Ibid., 113.
57. Ibid., 121.
58. Ibid., 126.
59. Imre Lakatos, "Falsification and the Methodology of Scientific Research Programmes," in *Criticism and the Growth of Knowledge*, ed. Imre Lakatos and Alan Musgrave (Cambridge: Cambridge University Press, 1970), 118.
60. Feenberg, *Questioning Technology*, 224.
61. Heidegger, *The Question*, 28 and 34.

CHAPTER FOUR

IAIN THOMSON

What's Wrong with Being a Technological Essentialist?

A Response to Feenberg

Introduction

Questioning Technology is Andrew Feenberg's third major work on the critical theory of technology in a decade, and it confirms his place as one of the world's leading philosophers of technology.[1] In an earlier examination of this important text, I traced out some of the philosophical and political tensions in the legacy of technology critique leading from Heidegger through Marcuse to Feenberg, and concluded that the critical theory of technology Feenberg elaborates in *Questioning Technology* remains much more conceptually indebted to Heidegger than Feenberg's own Marcuseanism had allowed him to admit. In response, Feenberg forthrightly acknowledged Heidegger's great influence on his work, but then went on to stress what he took to be the most important outstanding difference between his own critical theory of technology and Heidegger's critique of our technological understanding of Being, namely, Heidegger's "untenable" *technological essentialism*.[2]

I would like to follow up on our previous exchange here by asking, What is at stake in Feenberg's claim that Heidegger is a technological essentialist? I pursue this question not only in order to vindicate much of Heidegger's groundbreaking ontological approach to the philosophy of technology, but also to clarify Feenberg's conceptual cartography of technological essentialism. Doing so, I believe, will help orient the approach of future philosophers of technology to one of its central theoretical controversies.

Technological Essentialism

In our previous debate, the fundamental philosophical difference between Heidegger's and Feenberg's understandings of technology emerged in deceptively stark terms. Feenberg argued that Heidegger's ontological understanding of technology is untenably essentialistic, while I maintained that "Feenberg's reading is never so hermeneutically violent as when he accuses Heidegger of being a technological 'essentialist.'" On closer inspection, however, things are not quite so simple; as we will see, technological essentialism turns out to be an extremely complex notion.[3] Indeed, if we are to evaluate Feenberg's critique of Heidegger, the first thing we need to do is establish the criteria that determine what counts as technological essentialism. To minimize potential objections, I will stick to the criteria set forth by Feenberg himself.

The necessary criterion seems obvious; to be a technological essentialist, one needs to believe that technology has an essence. This criterion is not sufficient for our purposes, however, because it does not tell us what makes technological essentialism *objectionable*. A radical constructivist such as Baudrillard or Latour might maintain that there is no technology, only particular technologies, and thus that *all* technological essentialisms are unsound; but whether or not this is a coherent position, it is clearly not one that Feenberg shares.[4] Feenberg proposes his own "theory of the essence of technology,"[5] so the mere belief that technology has an essence cannot be sufficient to qualify one as the kind of technological essentialist to whom Feenberg objects. Thus, despite Feenberg's rather incautious claim that "[t]he basic problem is essentialism,"[6] it seems that the problem is not with technological essentialism as such, but rather with particular *kinds* of technological essentialism.

In fact, if I understand him correctly, Feenberg objects to technological essentialists such as Heidegger, Ellul, Borgmann, and Habermas because each commits himself to at least one of three particular claims about the essence of technology, claims that render their technological essentialisms unacceptable: *ahistoricism*, *substantivism*, and *one-dimensionalism*. Our next task will be to unpack these three essentialist claims with the goal of understanding what they are and why they are objectionable. We will then come back to each claim in turn and ask whether Heidegger holds any of the objectionable doctrines in question.

Ahistoricism

What is *ahistorical* technological essentialism, and what is wrong with it? According to Feenberg, an ahistorical technological essentialist is someone who interprets the "historically specific phenomenon [of technology] in terms of a transhistorical conceptual construction."[7] Thus, for example, Weber and Habermas understand the essence of technology in terms of "rational control [and] efficiency,"[8] while Heidegger understands it as the reduction of "everything to functions and raw materials."[9] What does Feenberg think is illegitimate about this? The problem is that, in an attempt to "fix the historical flux [of technology] in a singular essence," ahistorical essentialists abstract their understandings of the essence of technology from the "socially and historically specific context" in which particular technologies are always embedded.[10] As a result, not only do these ahistoricist theories fail to understand "the essence of technology as a social phenomenon,"[11] but their complete abstraction from sociohistorical context yields an "essentially unhistorical" understanding of the essence of technology, which is "no longer credible"[12] and so needs to be replaced by Feenberg's own "historical concept of essence."[13]

We will hold off on evaluating this objection and asking whether or not it really applies to Heidegger until the two other objectionable forms of technological essentialism are on the table.

Substantivism

Let us turn, then, to substantivism, the second form of technological essentialism Feenberg seeks to vitiate and surpass. What is *substantivist* essentialism, and what is wrong with it? Feenberg characterizes substantivism as the claim that the essence of technology comes from beyond us and is thus out of our control. Substantivists from Marx to Heidegger understand technology as "an autonomous force separate from society, . . .impinging on social life from the alien realm of reason."[14] For the substantivist, the essence of technology seems to be shaping history from outside, imposing itself as though from a metaphysical beyond that entirely escapes human control. We can easily understand why Feenberg finds substantivism so objectionable if we remember that he is a critical theorist who believes that "[t]he fundamental problem of democracy today" is the question of how to "ensure the survival of agency in this increasingly technological universe."[15] The substantivist's belief that the essence of technology is beyond human control seems to entail a fatalistic attitude about the historical impact of technology, a fatalism that runs directly counter to Feenberg's attempt to preserve a meaningful sense of agency in our increasingly technological world.

One-Dimensionalism

Finally, Feenberg objects to those technological essentialists who subscribe to what he calls one-dimensional thinking, the belief that all technological devices express the same essence.[16] What is wrong with claiming that the myriad diversity of technological devices all express a common essence? The problem, Feenberg contends, is that one-dimensional technological essentialists must either reject or embrace technology whole cloth. There is no room within one-dimensional conceptions of technology for a fine-grained analysis capable of appreciating both the positive potentials and the deleterious effects of the ever more pervasive rule of technology in our everyday lives. For the critical theorist of technology, an uncritical embrace of the totality of technological devices is just as unsound as a technophobic rejection of technology *tout court*.

In sum, then, Feenberg's objections go not to technological essentialism as such, but rather to three specific kinds of technological essentialism: the ahistoricisms, which illegitimately elide technology's embeddedness within sociohistorical currents that continue to shape it, the substantivisms that adopt a politically dangerous fatalism by viewing technology as a force completely beyond our control, and the one-dimensionalisms, which treat all technological devices as of a kind and thereby preclude any balanced critique of technology's benefits as well as its harms. With these three objectionable varieties of technological essentialism laid out before us, we are ready to evaluate Feenberg's critique of Heidegger's technological essentialism. So let us ask: Is Heidegger's conception of the essence of technology unacceptably ahistorical, substantivist, or one-dimensional?

Heidegger on the Essence of Technology

What exactly is Heidegger's understanding of the essence of technology? Heidegger's most famous claim, that the essence of technology is nothing technological, may not initially seem to be of much help. But as I explained in our earlier debate, "essence" is an important term of art for Heidegger, a term he painstakingly explains in his famous 1955 essay on "The Question Concerning Technology." Drawing on these careful remarks, I argued that Heidegger's claim that "the essence of technology is nothing technological" is best approached in terms of the paradox of the measure: height is not high, treeness is not itself a tree, and the essence of technology is nothing technological. If we want to understand the "essence of technology," Heidegger contends, we cannot conceive of "essence" the way we have been doing since Plato (as what "*permanently* endures"), for that makes it seem as if "by the [essence of] technology we mean some mythological abstraction." Instead, we need to think of "essence" as a verb, as the way in which things "essence" (*west*) or "remain in play" (*im Spiel bleibt*).[17] So conceived, "the essence of technology" denotes the way in which intelligibility *happens* for us these days. In short, the referent of the phrase "the essence of technology" is our current constellation of intelligibility, "enframing" (*das Gestell*), the historical "mode of revealing" in which things increasingly show up only as resources to be optimized.

According to Heidegger, enframing is grounded in our metaphysical understanding of what-is, an "ontotheology" transmitted to us by Nietzsche. In Heidegger's history of Being, the great metaphysicians articulate and disseminate an understanding of what beings *are*, and in so doing establish the most basic conceptual parameters and standards of legitimacy for each historical epoch of intelligibility. These metaphysicians' ontotheologies function historically like self-fulfilling prophecies, reshaping intelligibility from the ground up. Nietzsche, on Heidegger's reading, understood the totality of what-is as eternally recurring will-to-power, an unending disaggregation and reaggregation of forces without purpose or goal. Now, our Western culture's unthinking reliance on this nihilistic Nietzschean ontotheology is leading us to transform all beings, ourselves included, into resources to be optimized and disposed of with maximal efficiency.[18]

As I explained in my earlier piece, Heidegger is deeply worried that within our current technological constellation of intelligibility, the post-Nietzschean epoch of enframing "[o]nly what is calculable in advance counts as being." Our technological understanding of being produces a "calculative thinking" which quantifies all qualitative relations, reducing all entities to bivalent, programmable "information," digitized data, which increasingly enters into what Baudrillard calls "a state of pure circulation."[19] As this historical transformation of beings into resources becomes more pervasive, it comes to elude our critical gaze; indeed, we begin to treat ourselves in the very terms that underlie our technological refashioning of the world: no longer as conscious subjects in an objective world but merely as resources to be optimized, ordered, and enhanced with maximal efficiency (whether cosmetically, psychopharmacologically, genetically, or even cybernetically). With this brief recapitulation in mind, let us now evaluate Feenberg's objections.

Ahistoricism?

First, ahistorical essentialism. Feenberg alleges that Heidegger's "ontologizing approach" to the history of technology entirely "cancels the historical dimension of his theory."[20] This objection seems to me to be the least plausible of the three. It is true that Heidegger understands technology ontologically, but he understands ontology historically. Remember that for Heidegger, the essence of technology is nothing other than an ontological self-understanding that has been repeatedly contested and redefined for the last twenty-five hundred years. This is why I contended in my earlier piece that Heidegger's historical understanding of the "essence" of technology may actually put his position closer to the "constructivist" than the "essentialist" camp, and it becomes clear that Feenberg shares a similar view when he advocates "a historical concept of essence" in *Questioning Technology*'s concluding chapter.[21] It was Heidegger who gave us the first historical conception of the essence of technology, and I think Feenberg should acknowledge this important conceptual debt while continuing to build on this tradition, rather than seeking to distance himself from Heidegger where there are no good philosophical reasons for doing so.

If this is right, how can Feenberg possibly think that Heidegger has an ahistorical conception of technology? It is instructive to pinpoint just where his reading goes wrong. Critics such as Derrida have long questioned Heidegger's epochal account of the history of Being. They were not persuaded by the way in which Heidegger's account divides the history of our ontological self-understanding into a series of unified constellations of intelligibility. Where Heidegger sees a series of overlapping but relatively distinct and durable ontological epochs, his critics claimed to observe a much greater degree of ontohistorical flux. Feenberg too questions the "periodization" of Heidegger's history of Being,[22] but his objection is more precise. In order to "deny all [historical] continuity and treat modern technology as unique,"[23] Heidegger introduces an untenably "sharp ontological break"[24] between modern technology and premodern craft. I contend that Heidegger does indeed claim that our contemporary technological understanding of Being is unique, but that he does not deny all historical continuity in order to make this point.[25]

If we understand, as too few commentators do, what exactly Heidegger thinks is unique about our contemporary historical self-understanding, then it becomes clear that Feenberg has bought into a widespread misreading when he attributes to Heidegger the "unconvincing" claim that the contemporary age is "uniquely oriented toward control."[26] According to Heidegger's understanding of enframing, the ontological "reduction to raw materials" is *not* "in the interests of control."[27] Why not? Because in our post-Nietzschean age there is increasingly no subject left to be doing the controlling. The subject too is being "sucked-up" into "the standing reserve"![28] This unprecedented absorption of the subject into the resource pool makes our contemporary world unique in Heidegger's eyes, but he still explains this ongoing development historically; put simply, it results from the fact that we postmoderns have turned the practices developed by the moderns for objectifying and controlling nature back *onto ourselves*.[29]

In fact, despite this misreading of Heidegger, Feenberg now seems to have taken the basic Heideggerian point on board. In a recent essay on "Modernity Theory and Technology Studies," Feenberg observes with grim irony that: "Modern societies are unique in de-worlding human beings in order to subject them to technical action—we call it management." As Feenberg here seems to recognize, Heidegger presciently described an alarming ontological trend which now appears disconnected from our actual sociohistorical reality only to those who are not paying attention.[30] It should be clear, then, that Heidegger's technological essentialism does not suffer from the ahistoricism Feenberg attributes to it. Let us turn to one of Feenberg's more telling objections, his claim that Heidegger's understanding of technology suffers from a politically dehabilitating substantivism.

Substantivism?

Earlier we saw that Feenberg is moved to reject technological substantivism, the belief that the essence of technology is outside of human control, because of the politically dangerous fatalism this seems to entail.[31] Of course, a philosopher cannot reject a philosophical doctrine solely because of its political consequences. Distressing political implications should lead us to subject a philosophical doctrine to especially relentless critical scrutiny, but ultimately such philosophical scrutiny must seek to determine whether or not the doctrine in question is true. And if a philosophical doctrine turns out to be true, then either we have to accept its political consequences, however disturbing, or else we have to work politically to bring about a change in the world that would subsequently falsify the doctrine.

The problem with Heidegger's substantivism, as Feenberg presents it, is that the truth of the doctrine would seem to preclude the latter, activist option. For if Heidegger's substantivism is right that it is simply not within our power to transform the essence of technology, then neither can we change the world so as subsequently to gain control over the essence of technology.[32] In fact, if Feenberg were correct about Heidegger's substantivism, this would place us before a strict aporia, since Heidegger recognizes that we cannot stop trying to take control of the essence of technology; the endeavor may be impossible, but it is also unavoidable. As enframers, "the drive to control everything is precisely what we do not control."[33] Nevertheless, "this is a situation about which something can be done—at least indirectly."[34] This caveat, which allows for the possibility that our actions could *indirectly* transform the essence of technology, is crucial, it seems to me, for vindicating Heidegger's substantivism against Feenberg's objection.

For Feenberg is right that if Heidegger thought we had no hope of ever transcending our technological understanding of Being, his insights would lead only to fatalistic despair. Fortunately, Heidegger's position is more complex than this. Let us recall, with Dreyfus, that "Heidegger's concern is the human *distress* caused by the *technological understanding of Being*, rather than the *destruction* caused by *specific technologies*." Heidegger thus approaches technology not as "a *problem* for which we must find a *solution* [which would be a technological approach], but [as] an *ontological condition* that requires a *transformation of our understanding of Being*."[35] From the Heideggerian perspective, then, the most profound philosophical difference between Feenberg and Heidegger concerns the level at which each pitches his critique of technology; Feenberg's strategy for responding to the problems associated with the increasing rule of technocracy takes place primarily at what Heidegger would call the "ontic" level. The problem with Feenberg's strategy is that our everyday ontic actions and decisions almost always take place within the fundamental conceptual parameters set for us by our current ontology, otherwise these actions would not make sense to ourselves or to others.

For those of us seeking to synthesize Heidegger's and Feenberg's powerful critiques of technology, the crucial question is: Can ontic political decisions and resistances of the type Feenberg puts his faith in ever effect the kind of ontological change Heidegger seeks? Ontologically, Heidegger is more of a realist than a constructivist; our understanding of what-is is something to which we are fundamentally *receptive*. We cannot simply legislate a new ontology. As Dreyfus nicely puts it, "A new sense of reality is not something that can be made the goal of a crash program like the moon flight."[36] But does Heidegger deny that our ontic decisions could ever build up enough steam to effect an ontological transformation? No; in fact, Heidegger explicitly recognized this possibility. As he wrote in the late 1930s:

> "World-historical" events are capable of assuming a scale never seen before. [The unprecedented magnitude of these events] at first speaks only to the rising frenzy in the unbounded domain of machination and numbers. It never speaks immediately for the emergence of essential decisions. But when, within these "world historical" events, a coming-together of the people sets itself up—and partly establishes the people's existence according to the style of these events—could not a pathway open here into the nearness of decision? Certainly, but with the supreme danger that the domain of this decision will be missed completely.[37]

In other words, it is possible that a confluence of ontic political struggles could open the space for a reconfiguration of our ontological self-understanding, but only if we are aware of the true radicality of that endeavor, the fact that it requires a fundamental transformation in the nature of our existence, not merely the redistribution of power or the realignment of particular interests.

As Dreyfus's famous Woodstock example is meant to show, it is possible that practices marginalized by our technological understanding of Being could become central to our self-understanding, radically transforming our sense of what is and what matters.[38] As I pointed out last time, although Feenberg's own project in clearly inspired by the Paris events of May '68 in which he participated, he is extremely wary of this revolutionary aspect in Heidegger's thinking because of the political direction it took Heidegger himself. But how different are Feenberg and Heidegger on this point? Do we not have Feenberg's own position if we simply replace Heidegger's politically dangerous Nietzschean-Wagnerian hope for a revolutionary *Gesamtkunstwerk*, a work of art that would transform our entire ontological self-understanding in one fell swoop, with the more modest hope that a "convergence" of differently situated political micro-struggles could evolve into a counterhegemony capable of permanently subverting our contemporary technocracy?[39]

If Heidegger steadfastly advocates the goal of ontological transformation, while Feenberg seeks to "reverse-engineer" a possible means to achieving this goal (through a confluence of "democratizing" ontic struggles over technological design), this should also lead us to wonder, I think, how much Heidegger and Feenberg really differ on the truth of substantivism. In our previous debate, I argued that Feenberg actually wavers back and forth on the substantivism question; this tension in Feenberg's view reflects a fundamental difference between the Marcusean and Heideggerian positions he has synthesized. He vacillates between a voluntaristic, Marcusean, May '68, "Progress will be what we want it to be" view, which exalts the human capacity to control our future through strategic interventions in the design process,[40] and a more substantivist Heideggerian view that suggests that while we cannot directly *control* the historical direction in which technology is taking us, we can nevertheless impact the future in small ways by learning to recognize, encourage, and support technological democratizations when they occur, while hoping that these ontic political interventions might yet indirectly foster an ontological transformation. The Marcusean position has the surface appeal of all heroic existential voluntarisms, but it ignores the very issue that led Heidegger to develop his ontological approach, indeed the very reason that Marcuse discipled himself to Heidegger before the war: however important, democratization without a corresponding ontological transformation will just end up replicating and reifying the technological understanding of Being.

Another thing this shows, I think, is that Feenberg's projected democratization of technological design needs to be supplemented by a pedagogical project aimed at the level of what the Greeks called *paideia*, the Germans *Bildung*, that is, an educational formation geared toward recognizing and encouraging the development of certain specific world-disclosing skills—one species of which would be those skills necessary for making appropriate democratizing interventions in the design process.[41] I will try to say a bit more about what sort of skills this pedagogical project should seek to inculcate as we evaluate Feenberg's final objection.

One-Dimensionalism?

Is Heidegger's technological essentialism one-dimensional? Does he believe that all technological devices express the same essence? In "The Question Concerning Technology," Heidegger explicitly denies that "enframing, the essence of technology" is "the common genus of everything technological." That is, in seeking to understand the essence of technology, Heidegger is not trying to fix the extension of the term; he is not seeking to determine what is and what is not a member of the class of technological devices.[42] Thus, he does not conceptualize technology's essence in terms of the commonalities shared by the hydroelectric plant, the autobahn, the cellular phone, the Internet, etc., the way a Platonist might conceive of the essence of trees as the genus uniting "oaks, beeches, birches, and firs."[43] Strictly speaking, then, Heidegger's understanding of the essence of technology is orthogonal to the question of whether or not all technological devices express the same essence.

Nevertheless, the question of whether Heidegger is a technological one-dimensionalist remains. And the answer, I think, is a qualified yes. Why? Because, as we have seen, Heidegger holds that *the essence of technology is nothing less than the ontological self-understanding of the age*. Insofar as we implicitly adopt the ontology of enframing, *everything* in the contemporary world will show up for us as reflecting the essence of technology, technological devices included. In this sense, then, Heidegger does seem to be a kind of technological one-dimensionalist. But do the negative consequences Feenberg attaches to this position obtain in Heidegger's case? Not unless Heidegger's understanding of the essence of technology forces him globally to reject technology. This, then, is the crucial question: Does Heidegger's one-dimensionalism force him to reject technology in toto?

Now, Heidegger is obviously no fan of technology; he seems, for instance, to have had a kind of visceral reaction to the sight of his neighbors "chained hourly and daily to their television" sets.[44] But even on the personal level, Heidegger seems occasionally to have been capable of distinguishing between those technological applications that serve, and those that undermine, the cause of phenomenology, the endeavor to go "To the things themselves!" For example, while watching a television show a friend put together to showcase the art of Paul Klee, Heidegger was appalled by the way the television moved over the paintings randomly and forced the eye away from one piece and on to the next prematurely, "hindering an intensive, quiet viewing as well as a lingering reflection, which each single work and the relations within it deserve." On the other hand, Heidegger deeply appreciated the way a televised soccer match revealed its subject, raving publicly that it showcased the "brilliance" of Franz Beckenbauer.[45] Of course, such anecdotes do not get us to the crux of the issue. For, however "technophobic"[46] Heidegger may have been personally, it is obvious to careful readers of his work that he does not advocate any monolithic rejection of technology *philosophically*. This should not be too surprising, since the philosophical implications of Heidegger's thinking often far exceed the rather narrow conclusions he himself drew from them.

In our previous debate, I reminded Feenberg of Heidegger's phenomenological description of the massive freeway interchange on the autobahn. Here in 1951, Heidegger treats the autobahn in terms of what he calls a "thing thinging," that is, as a work of art reflecting back to us the ontological self-understanding of the age.[47] In response, Feenberg acknowledged that in these passages on the autobahn bridge, "Heidegger discusses modern technology without negativism or nostalgia and suggests an innovative approach to understanding it." Nevertheless, Feenberg countered, Heidegger's "defenders have to admit that the famous highway bridge passage is the one and only instance in his whole corpus of a positive evaluation of modern technology." Feenberg may well be right about this; Heidegger's brief phenomenological meditation on the autobahn interchange as a paradigm reflecting our ontological self-understanding may be the only "positive evaluation of modern technology" to be found in his published work. But is not this single, carefully thought-out exception sufficient to prove that Heidegger does not reject technology whole cloth?

In his meditation on the autobahn interchange, Heidegger's concern is not to valorize this technological paradigm, but rather to help us recognize that, as the Internet now makes plain, we are increasingly treating our world and ourselves as a kind of "network of long distance traffic, paced as calculated for maximum yield."[48] Indeed, the only thing making this a "positive evaluation" (as Feenberg puts it) is the fact that, in his phenomenological description of the autobahn interchange, Heidegger is attempting to get us to notice the presence of "the divinities" that linger in the background of even our most advanced technological constructions.[49] When he refers to the presence of the divine, Heidegger is evoking those meanings that cannot be explained solely in terms of human will, encouraging us to attend to that preconceptual phenomenological "presencing" upon which all of our interpretations rest, a "presencing" that Heidegger thinks will be a prime source of any "new paradigm . . . rich enough and resistant enough to give a new meaningful direction to our lives."[50]

Like his meditation on the place of "earth" in the work of art, Heidegger's resacralization of the simple "thing" reminds us that the conditioned has its roots in the unconditioned, the secular in the sacred, and thus suggests that we should adopt a very different attitude toward our world, a *Grundstimmung* much more reflective and thankful than the thoroughgoing instrumental reasoning characteristic of our technological mode of revealing. Indeed, as Dreyfus has argued, Heidegger is convinced that we should be *grateful* for the essence of technology; for without this cultural clearing, "nothing would show up *as* anything at all, and no possibilities for action would make sense."[51] To recognize enframing as our current constellation of intelligibility is to recognize our ontological receptivity in addition to our active role as disclosers of what-is. If we can incorporate a sense of this receptive spontaneity into our practices, we can learn to relate to things with a phenomenological comportment open to alterity and difference (on the ontological as well as the more fashionable ontic level), a comportment through which Heidegger believes we may yet disclose the constituent elements of a post-technological ontology.

This may sound mysterious, but in his 1949 essay on "The Turning," Heidegger unequivocally states that he is not advocating anything as ridiculous as the abandonment of technology. In the postnihilistic future that Heidegger worked philosophically to help envision and achieve, "Technology," he repeats, "will not be done away with. Technology will not be struck down, and certainly it will not be destroyed." Indeed, Heidegger can no longer be confused with a Luddite longing for a nostalgic return to a pretechnological society; in his final interview (given in 1966), he reiterates that the technological world must be "transcended, in the Hegelian sense [that is, incorporated at a higher level], not pushed aside."[52] Heidegger's critics may object that he does not provide enough guidance about how practicing an open phenomenological comportment will allow us to transcend our current technological understanding of Being, but he cannot be accused of a reactionary rejection of technological devices, and even less of wanting to reject the essence of technology, which, he says, would be madness, "a desire to unhinge the essence of humanity."[53]

One further point is clear; Heidegger did not believe that our technological understanding of Being could be transcended through a phenomenological practice disconnected from sociohistorical reality. It will doubtless surprise those who have been taken in by a one-sided stereotype to hear that when Heidegger was devoting a great deal of thought to the question of the relation between "the work of art and the power plant," he spent "several days visiting power plants under the direction of professors from technical colleges."[54] The fruits of such phenomenological labors are undeniable. As I noted previously, when Heidegger looked out at the autobahn interchange and the power plant on the Ister and found words that powerfully describe those fundamental transformations in our self-understanding which are only now becoming obvious with the advent of the Internet, word processing, genetic research, and cloning, his was not what Auden called "The dazed uncomprehending stare / Of the Danubian despair."[55]

Conclusion

In sum, then, Heidegger appears to be a technological essentialist, but of a largely unobjectionable variety. For as we have seen, he rejects ahistoricism entirely, and the forms of one-dimensionalism and substantivism he accepts lack these doctrines' usual negative implications. Indeed, Heidegger's substantivism offers an indirect response to Feenberg's political objection, a response that rests on a much more thorough philosophical analysis than the voluntaristically motivated objection, and Heidegger's one-dimensionalism clearly does not force him into any global rejection of technology. Heidegger's rather limited technological essentialism thus does little to discredit his profound ontological understanding of the historical impact of technology. Indeed, even where Feenberg's rhetoric conceals this fact, his important critical theory of technology has obviously learned a great deal from the ontological and phenomenological subtleties found in Heidegger's work, and there is every reason to suppose that Feenberg and future philosophers of technology will continue to find in Heidegger's reflections a challenging and rewarding source of philosophical inspiration.[56]

Notes

Reprinted from "What's Wrong with Being a Technological Essentialist? A Response to Feenberg," *Inquiry* 43 (2000): pp. 429–44 (www.tandf.no/inquiry) by permission of Taylor and Francis AS.

1. See Feenberg, *Critical Theory of Technology* (New York and Oxford: Oxford University Press, 1991); *Alternative Modernity: The Technical Turn in Philosophy and Social Theory* (Berkeley: University of California Press, 1995); and *Questioning Technology* (London and New York: Routledge, 1999). Unprefixed page references throughout refer to this last work.
2. See my "From the Question Concerning Technology to the Quest for a Democratic Technology: Heidegger, Marcuse, Feenberg," and Feenberg's "Response to Critics," *Inquiry* 43 (2000): 203–15, 225–37.
3. When Feenberg criticizes technological essentialism, he is not thinking of the Kripkean claim that an essence is a property a thing possesses necessarily. (See Saul Kripke, *Naming and Necessity* [Cambridge: Harvard University Press, 1980].) He is simply using essentialism as a descriptive term to characterize a fairly wide range of theories about technology with which he disagrees.

4. It is not clear that the radical constructivists' sloganistic claim—that there is no technology, only technologies—makes sense; in virtue of what are all these different technologies "technologies"? There are, of course, other affinities between Feenberg and the constructivist camp (see esp. 83–85).
5. Feenberg, *Questioning*, 17.
6. Ibid.
7. Ibid., 15.
8. Ibid., vii.
9. Ibid., viii.
10. Ibid., 17.
11. Ibid.
12. Ibid., 15.
13. Ibid., 201.
14. Ibid., vii.
15. Ibid., 101.
16. Feenberg appropriates this term from Marcuse, then applies it back to Marcuse's own "one-dimensional" conception of our "fully administered" society.
17. Heidegger, *The Question Concerning Technology*, trans. W. Lovitt (New York: Harper and Row, 1977), 4, 30–31.
18. I develop and defend these ideas in "Ontotheology: Understanding Heidegger's *Destruktion* of Metaphysics," *International Journal of Philosophical Studies* 8, no. 3 (2000): 297–327.
19. Martin Heidegger, "Traditional Language and Technological Language," trans. W. Gregory, *Journal of Philosophical Research* XXIII (1998): 136. Heidegger, *Discourse on Thinking*, trans. J. Anderson and E. Freund (New York: Harper and Row, 1966), 46; Heidegger, "Traditional Language and Technological Language," 139; Jean Baudrillard, *The Transparency of Evil*, trans. J. Benedict (London: Verso, 1993), 4. For Heidegger, the "quantitative dominates all beings," and when this "limitless "quantification"" exhausts all qualitative relations, we come to treat "quantity *as* quality." See Heidegger, *Contributions to Philosophy (From Enowning)*, trans. P. Emad and K. Maly (Bloomington: Indiana University Press, 1999), 95/*Beiträge zur Philosophie (Vom Ereignis)*, *Gesamtausgabe* vol. 65 (ed.) F.-W. von Hermann (Frankfurt a.M.: Vittorio Klostermann, 1989), [hereafter GA65], 137; ibid., 94 (my emphasis)/GA65, 135.
20. Feenberg, *Questioning*, 16.
21. Ibid., 201.
22. Ibid., 15.
23. Ibid.
24. Ibid., 16.
25. Glazebrook shows this clearly in "From *Phusis* to Nature, *Technê* to Technology: Heidegger on Aristotle, Galileo, and Newton," *The Southern Journal of Philosophy* XXXVIII (2000).
26. Feenberg, *Questioning*, 15.
27. Ibid., 178.
28. See Heidegger, "Science and Reflection," *The Question Concerning Technology*, op. cit., 173. In Feenberg's recent "Modernity Theory and Technology Studies: Reflections on Bridging the Gap" (unpublished, available on Feenberg's Web page), he again attributes to Heidegger "the familiar complaint about modernity's obsession with efficiency and control." Of course,

Feenberg would be right if he were distinguishing "modernity" from "postmodernity," rather than using modernity to designate the contemporary age, as he does here.

29. Heidegger's claim is that when modern subjects dominating an objective world begin to transform themselves into objects, the subject/object distinction itself is undermined, and these subjects thus put themselves on the path toward becoming just one more resource to be *optimized*, i.e., "secured and ordered *for the sake of flexible use*." See Charles Spinosa, Fernando Flores, and Hubert L. Dreyfus, "Skills, Historical Disclosing, and the End of History: A Response to Our Critics," *Inquiry* 38. no. 1–2 (1995): 188 (my emphasis).
30. The passage from modernity to postmodernity was, for Heidegger, already clearly visible in the transformation of employment agencies into "human resource" departments. (See 1955's *The Question Concerning Technology*, 18.) Our contemporary reduction of teachers and scholars to on-line "content providers" merely extends—and so clarifies—the logic whereby modern subjects become postmodern resources, a logic that (as we have seen) Heidegger traces philosophically back to Nietzsche's metaphysics.
31. According to Feenberg, Heidegger is fatalistic because he ignores the bottom-up perspective of those "enrolled" within technological networks and so misses their "subjugated wisdom": technologies can be appropriated from below, diverted away from the fixed ends for which they were originally designed. But Heidegger would not deny that specific technological designs can be subverted in this way. The crucial question is whether such "ontic" subversions could ever culminate in an ontological transcendence of the technological mode of revealing. As I show below, Heidegger did believe in just such a possibility, but he did not believe that it could be accomplished simply by steering the course of technological development from within. Thus, in 1940, Heidegger bemoans our contemporary age's call for the Nietzschean *Übermensch*, our sense that "[w]hat is needed is a form of mankind that is from top to bottom equal to the unique fundamental essence of contemporary technology and its metaphysical truth; that is to say, that lets itself be entirely dominated by the essence of technology precisely in order to steer and deploy individual technological processes and possibilities." Feenberg himself can be understood as advocating precisely this voluntaristic, Nietzschean strategy. See Heidegger, *Nietzsche*, *Volume Four: Nihilism*, ed. David Krell, trans. F. A. Capuzzi (San Francisco: Harper and Row, 1982), 117.
32. If substantivism is right that we cannot control the essence of technology (and clearly this is meant as the time-independent claim that the essence of technology is out of our control now and forever—otherwise it would not be objectionable), then there is no non–question begging way to say that we could change the world such that we *could* control the essence of technology.
33. See Dreyfus, "Heidegger on the Connection Between Nihilism, Art, Technology, and Politics," in *The Cambridge Companion to Heidegger*, ed. C. Guignon (Cambridge: Cambridge University Press, 1993), 307–10. On Heidegger's alleged "fatalism," see also Young, *Heidegger, Philosophy, Nazism*, 188–91.
34. Dreyfus, "Nihilism, Art, Technology, and Politics," 305.
35. Ibid.
36. Ibid., 310. *Pace* Winograd and Flores, then, we are not ontological designers. We are, rather, ontic designers. See Terry Winnograd and Fernando Flores,

Understanding Computers and Cognition: A New Foundation for Design (Reading, MA: Addison-Wesley, 1986).

37. Heidegger, *Contributions to Philosophy*, 68/GA65, 98. The context of this passage is philosophically and politically problematic: philosophically, because here Heidegger is still naively committed to the metaphysical project of establishing a new historical ground for beings (by "deciding" a new understanding of the Being of beings); politically, because Heidegger not only connects this metaphysical project with the "people" (*Volk*), but even asserts the "singularity" of this folk's "origin and mission," grounding this "destiny" in "the singularity of Be-ing itself" (ibid., 67/ 97). This nationalistic philosophical appropriation of the Jewish trope of the chosen people, sometime between 1936–37, is especially troubling. Nevertheless, Heidegger does not assert the heterogeneity of the ontic and ontological domains, as Feenberg seems to believe.
38. See my and see Dreyfus, "Nihilism, Art, Technology, and Politics," 311. *Cf.* Hubert L. Dreyfus, "Heidegger on Gaining a Free Relationship to Technology," in *Technology and the Politics of Knowledge*, ed. Andrew Feenberg and Alastair Hannay (Bloomington: Indiana University Press, 1995), 106.
39. There are, of course, important differences between the revolutionary and evolutionary perspectives. Indeed, Heidegger's own adoption of the revolutionary view seems to have desensitized him to the real human suffering ushered in by the pseudo-revolution of 1933. Nevertheless, Heidegger's critique of the evolutionary view is right about at least this much: the mere fact that the hands of the clocks keep turning, so to speak, does not mean that history is moving toward any sort of ontological transformation.
40. Feenberg, *Questioning*, 22.
41. Charles Spinosa, Fernando Flores, and Hubert L. Dreyfus's groundbreaking work, *Disclosing New Worlds: Entrepreneurship, Democratic Action, and the Cultivation of Solidarity* (Cambridge: The MIT Press, 1997) closes by issuing a similar call (see esp. pp. 171–73), and Feenberg has recently recognized this affinity in his "Modernity Theory and Technology Studies." For a sympathetic explication of Heidegger's mature understanding of university education, see my "Heidegger on Ontological Education, or: How We Become What We Are," *Inquiry* 44. no. 2 (2001): 243–68.
42. See Heidegger, *The Question Concerning Technology*, 29. This, I take it, is what Dreyfus means when he says: "[W]hen he asks about the essence of technology we must understand that Heidegger is not seeking a definition. His question cannot be answered by defining our concept of technology." See "Nihilism, Art, Technology, and Politics," 305.
43. The Platonist conceives of the essence of the different species of trees in terms of the abstract idea of "treeness," but Heidegger does not analogously conceptualize the essence of the diversity of technological devices by abstracting toward a kind of "technicity" [*Technik*] or "machination" [*Machenschaft*]. Indeed, by 1938, he has recognized that "Machination itself . . . is the essential swaying of Being [*die Wesung des Seyns*]," i.e., that what technological devices share in common is their ontological mode of revealing (which is rooted in Nietzsche's metaphysics of "constant overcoming," his "ontotheology" of "eternally recurring will to power"). Thus, Heidegger writes: "The bewitchment by technicity and its constantly self-surpassing progress is only *one* sign of this enchantment, by which everything presses forth into calculation,

usage, breeding, manageability, and regulation." See Martin Heidegger, *Contributions to Philosophy*, 89/GA65, 128; ibid., 87/ 124.

44. Heidegger, *Discourse on Thinking*, 50. Thirty years earlier (in 1928), Heidegger pictured technology as rampaging across the globe "like a beast off its leash." See Heidegger, *The Metaphysical Foundations of Logic*, trans. M. Heim (Bloomington: Indiana University Press, 1984), 215.
45. See Heinrich W. Petzet, *Encounters and Dialogues with Martin Heidegger: 1929–1976*, trans. P. Emad and K. Maly (Chicago: University of Chicago Press, 1993), 149–50, 210.
46. Feenberg, *Questioning*, 151.
47. See Heidegger, *Poetry, Language, Thought*, trans. A. Hofstadter (New York: Harper and Row, 1971), 152–53. Put simply, Heidegger conceives of works of art on three orders of magnitude: micro-paradigms ("things") such as Van Gogh's painting of the peasant shoes; paradigms ("works of art" proper) such as the autobahn interchange; and macro-paradigms ("gods") such as the Greek temple. (See my "The Silence of the Limbs: Critiquing Culture from A Heideggerian Understanding of the Work of Art," *Enculturation* 2, no. 1 [1998]). At one end of the continuum, things *gather* a local world, at the other, artworks *reconfigure* the worlds they bring into focus (in the extreme case, the "god", inaugurating a new onto-historical epoch).
48. Heidegger, "Building Dwelling Thinking," *Poetry, Language, Thought*, 152.
49. Ibid., 153. For a fascinating analysis of freeway interchanges as artworks reflecting back the self-understanding of the age, see David Brodsly's monograph, *L.A. Freeway: An Appreciative Essay* (Berkeley: University of California Press, 1981).
50. Dreyfus, "Nihilism, Art, Technology, and Politics," 311. As possible sources of such a new paradigm, Dreyfus stresses those "marginal practices" that have not yet been completely "mobilized as resources," "such as friendship, backpacking in the wilderness, and drinking the local wine with friends" (310). I would add that for Heidegger a crucial role will be played by "presencing" (*Anwesen*), which I understand as a preconceptual phenomenological givenness and extraconceptual phenomenological excess that existing practices never exhaust.
51. Ibid., 307. See also Dreyfus, "Heidegger on Gaining a Free Relationship to Technology." I am indebted to Julian Young for the former point.
52. See Martin Heidegger, "The Turning," in *The Question Concerning Technology*, 38; Martin Heidegger, "The Spiegel Interview," trans. L. Harries, in *Martin Heidegger and National Socialism*, ed. G. Neske and E. Kettering (New York: Paragon House, 1990), 63. For Hegel, "transcending" (or "sublating," *Aufheben*) "is at once a negating and a preserving." See Hegel, *Phenomenology of Spirit*, trans. A. V. Miller (Oxford: Oxford University Press, 1977), 68. As Heidegger clearly explains, "Sublated does not mean done away with, but raised up, kept, and preserved in the new creation." See Heidegger, "Phenomenology and Theology," trans. J. G. Hart and J. C. Maraldo, in *Pathmarks*, ed. William McNeill (Cambridge: Cambridge University Press, 1998), 51.
53. See Heidegger, *Nietzsche, Volume Four: Nihilism*, 223. For more guidance about how Heideggerian "world disclosing" takes place concretely, see Spinosa, Flores, and Dreyfus, *Disclosing New Worlds*.
54. See Petzet, *Encounters and Dialogues*, 145–46.

55. For a persuasive argument to this effect (one that Feenberg does not yet seem to have taken the full measure of), see Hubert Dreyfus and Charles Spinosa, "Highway Bridges and Feasts: Heidegger and Borgmann on How to Affirm Technology," *Man and World* 30, no. 2 (1997).
56. An earlier version of this chapter was presented to the Western Division of the American Philosophical Association, Albuquerque, NM, April 6, 2000. I would like to thank Bert Dreyfus, Andy Feenberg, Jerry Doppelt, Adrian Cussins, Wayne Martin, and John Taber for their helpful comments and criticisms.

CHAPTER FIVE

LARRY A. HICKMAN

From Critical Theory to Pragmatism

Feenberg's Progress

Over the course of more than two decades, during which he has published an impressive number of books and essays, Andrew Feenberg has established himself as an important representative of a new generation of critical theorists. Consistently insightful and articulate, he has developed a trenchant critique of technological culture that has taken as its point of departure the humanistic Marxism of his mentor Herbert Marcuse. In his recent book *Questioning Technology* (1999), he presents what is arguably his most successful attempt to date to construct a major revision of the critique of technology advanced by Marcuse and other "first generation" critical theorists, as well as by their "second generation" heirs, such as Habermas. At one level his work can be read as fulfilling a promise the details of which Marcuse just left vague. At a deeper level, however, *Questioning Technology* can be viewed as a move away from some of the core ideas of the earlier critical theorists, including Marcuse.

As a student of Marcuse, Feenberg might plausibly be thought to belong to the second generation of critical theory. Following Joel Anderson's excellent essay on the history of the Frankfurt School, however, our best option is to place Feenberg's published work squarely within critical theory's "third" generation. If the "first" generation was interested in emancipation from instrumental rationality as ideology by means of reflective social science, and if the "second" generation focused on the development of communicational tools to promote moral development and respect for constitutionality, as well as to overcome social pathologies such as extreme nationalism, xenophobia, and the colonization of the lifeworld by technoscientific rationality, then the "third" generation, whose experiences were formed by the events of 1968, has abandoned the essentialist and substantialist views of its forebears in favor of positions that are more thoroughly functionalist and constructivist. This generation has turned its attention to problems of pluralism, multiculturalism, and globalization, and has tended to view problems of technoscience not as separate from, but as a part of social life.[1]

To those familiar with the central ideas of American pragmatism, some of the planks in Feenberg's platform will therefore appear remarkably familiar. More specifically, Feenberg's revisions of Marcuse have the interesting effect of moving his critique noticeably in the direction of the instrumental version of pragmatism that was developed in the first half of the twentieth century by John Dewey. It is perhaps not surprising that this should have occurred, given attempts by second generation critical theorists Habermas and Apel to appropriate the insights of C. S. Peirce, and the influential studies of the work of G. H. Mead that have been published by Hans Joas, who might be regarded as a kind of third generation critical theorist.

But this similarity between the Feenberg of *Questioning Technology* and Dewey's pragmatic critique of technology, I suggest, is all the more significant given the fact that he, Feenberg, has not given his readers much evidence that he is aware of this situation. In *Questioning Technology*, for example, he devotes a total of about a half page to Dewey. On page 136 he discusses Dewey's treatment of democratic deliberation and then dismisses him as having exhibited an "uncritical confidence in science and technology."[2] Ten pages later, however, he reminds us that Dewey foresaw how "the dispersion of the technological citizenry" and other factors, including a "media-dominated public process," would account "for the passivity of a society which has not yet grasped how profoundly affected it is by technology."[3]

The apparent conflict between these two assessments may in fact not be so great as it at first appears. Although Dewey did in fact have a measure of confidence in science and technology, or what is now frequently termed "technoscience," his attempts to present a democratized critique of technology are remarkably similar to those that Feenberg himself is now advancing. For those who are sympathetic to the programs of the American pragmatists, as I myself am, what I have termed "Feenberg's Progress" is therefore a matter to be applauded. That is what I intend to do in this chapter.

It is probably best to begin my account by taking the measure of where Feenberg thinks we have been historically with respect to the philosophical critique of technology. Presenting "before and after" snapshots in the preface to *Questioning Technology*, he describes what he views as "a fateful change in our understanding of technology."[4] What is this change? Put simply, it is that prior to what he regards as the extensions of democracy with respect to the technical sphere that are only now coming to fruition, philosophical critiques of technology were universally "essentialist."

On the cultural and political Right, there were the romantics: the Ruskins and the Heideggers who viewed technology as the root cause of all that was dehumanizing. On the cultural and political Left there were socialists and progressives who tended to an uncritical acceptance of everything that came off the engineers' drawing boards. But it should not be thought that these opposing and frequently conflicting camps had nothing in common. As Feenberg views matters, they "all agreed that technology was an autonomous force separate from society, a kind of second nature impinging on social life from the alien realm of reason in which science too finds its source."[5]

It is worth noting that although first generation critical theorists Horkheimer and Adorno were constructivists with respect to what they regarded as "the social world," they also espoused various versions of essentialism when it came to technology (which they placed over against the social). Marcuse embraced this view in a slightly weaker form: although he never worked out the details, his view was apparently that political reform was a necessary, but not sufficient, condition for the reform of technology.[6] Moreover, it will hardly be news even to casual readers of second generation critical theorist Habermas that his early work advanced a version of technological essentialism—technoscience was identified as and reduced to "instrumental rationality" and a gathering of facts—that he has since neither rejected nor subjected to significant revision.

That is the "before" snapshot. Now, however, the situation has changed. Conditions of emerging and expanding democratization have led critics to question essentialism, as well as the more extreme view advanced by Jacques Ellul and others, namely that the essence of technology is nefarious through and through. For Ellul, technology is a thing: it is a debilitating, all-consuming, autonomous force. As a response to these new conditions, Feenberg thinks, it now seems important to move technology from the "autonomous/other" column into a column that might be labeled "our social matrix." Put another way, the idea of the essentialists that technology occupies a location separate from the places where meaning and value are constructed has given way to the notion that meaning and value are also constructed in the context of technoscientific decision making.

Feenberg's response to this changed situation has been to issue a manifesto. "The time has therefore come," he writes, "for an anti-essentialist philosophy of technology."[7] His book is dedicated to working through what he thinks will be the characteristics of this new anti-essentialist philosophy.

The new anti-essentialism will in his view be an invigorated constructivism that allows for the possibility of difference where there was formerly only the monolithic "technology as thing" and that therefore takes into account, as he puts it, the "social and historical specificity of technological systems, the relativity of technical design and use to the culture and strategies of a variety of technical actors."[8] As I have already indicated, first generation critical theorists Horkheimer and Adorno were also constructivists. But their constructivism was limited to what they regarded as the social world—a world that they perceived as existing over against the world of technoscience, in which instrumental rationality held sway, and also over against nature, which was treated as just "given."[9]

What does this mean in practice? For one thing, it means that Feenberg has adopted a more comprehensive and explicit constructivist position, which holds that technology is neither determining nor neutral.[10] Here Feenberg indicates that he is following the suggestion of Don Ihde, that "technology is what it is in some use-context."[11]

But of course this notion is instrumentalist as well as constructivist, and it has a fairly long pedigree. It recalls Marshall McLuhan's insight that media are the extensions of "man," and also what Melvin Kranzberg was fond of calling "Kranzberg's First Law,"[12] namely, that technology is neither positive, nor negative, nor neutral. Does this mean that the methods and devices of technology are vectorless, that they do not possess momentum? Of course not. The point asserted by these instrumentalists is just that even though technological artifacts often possess a high degree of momentum, even to a degree that it is sometimes almost impossible to divert or overcome their motion, there is nevertheless a relation of feedback between the selection of tools and methods and the influence that those tools and methods have over our lives. As McLuhan put it, we shape our tools; thereafter they shape us.

It is also worth noting that this type of instrumentalism/constructivism had already been well articulated long before it resurfaced during the 1960s in the work of media theorist McLuhan and historian Kranzberg. One version, for example was developed at the end of the nineteenth century and the beginning of the twentieth by John Dewey. The very name he gave to his version of pragmatism might well have served as a clue to sharp-eared philosophers of technology that they might find something of interest in his work: he called his view "instrumentalism." Given the long history of instrumentalism/constructivism as an approach to the philosophy of technology, therefore, Feenberg's manifesto—that "the time has therefore come for an anti-essentialist philosophy of technology"—has an odd (but welcome) ring to the ears of this pragmatist.

From the negative side, as a part of his rejection of the technological determinism that is evident in the work of Horkheimer and Adorno and lurks in the background of the work of Marcuse, as well as from the positive side, as a part of his emerging commitment to an invigorated constructivism that includes technoscience, Feenberg has argued that "technology is not just the rational control of nature; both its development and impact are intrinsically social."[13] Moreover, "technologies are not physical devices that can be extricated from contingent social values; the technical always already incorporates the social into its structure."[14]

He has further insisted that "this view undermines the customary reliance on efficiency as a criterion of technological development . . . [which] opens broad possibilities of change foreclosed by the usual understanding of technology."[15] If one reads this statement in the light of the work of Horkheimer, Adorno, and Marcuse, then "efficiency" should probably be understood as "instrumental reason" or what Langdon Winner has called "straight-line instrumentalism," and "the usual understanding of technology" should probably be read as the views espoused by the first and second generation critical theorists.

In short, Feenberg is attempting to break up the old monolithic models of technology and replace them with a model that places decision making with respect to tools, methods, and techniques squarely within the sphere of the normative/evaluative, or what some have called "the lifeworld." In his view, technology is not something foreign to human life. It is not, as his predecessors thought, ideology by another name. It is not an unrestrained quest for efficiency. It is not about the domination of nature.

What then *is* technology? It is a natural activity of human beings, a part of their attempt to secure transitory goods and to improve the conditions of their lives, both as individuals and as groups. It is a method of decision making in which means and ends are weighed against one another in a process of continual readjustment. It is complex, multifaceted, and, with the requisite amount of effort, even reversible. It is possible to speak of technological progress (and regress, as well). Given the factors that I have listed in this paragraph, it should be clear that Feenberg's break with the first generation critical theorists is massive, and it is dramatic. More important for the subject of this paper, however, this paragraph also contains a pretty good representation of John Dewey's pragmatic technology.

As I have already indicated, anyone who finds Dewey's pragmatist critique of technology attractive can only applaud Feenberg's Progress. The fact is that his position now resembles that of Dewey much more than that of his teacher Marcuse. Already in the 1890s Dewey was beginning to articulate a critique of technology that comprehends most of the elements that I just listed. He continued to refine that critique right up until his death in 1952.

First, like Feenberg, he viewed technology as a multifaceted enterprise. For Dewey, this meant that technology was more or less interdefinable with inquiry in the broad sense of the term. Dewey rejected essentialism early on, calling for a functionalized understanding of what philosophers from the Greeks to the moderns had called essences. According to his functionalist view, the essence of an event, object, or institution amounts to those aspects that we ("we" in this instance meaning members of various publics, or communities of inquiry) find of sufficient relevance to our own needs and interests that we select and utilize them to characterize that event, object, or institution.

Dewey claimed that essentialists tend to commit what he called "the philosophical fallacy," namely the taking of something that is the result of inquiry as if it had existed prior to that sequence of inquiry in precisely the form that the inquiry determined it to be. Dewey's "philosophical fallacy" is, of course, more or less what Whitehead would later call "the fallacy of misplaced concreteness," which is in turn the very fallacy that Feenberg condemns in the closing pages of *Questioning Technology*. When technology is stripped of its values and social context, he (Feenberg) writes, as sometimes occurs in engineering and management contexts, "technology emerges from this striptease as a pure instance of contrived causal interaction. To reduce technology to a device and the device to the laws of its operation is somehow obvious, but it is typical fallacy of misplaced concreteness."[16]

Like Feenberg, but unlike Marcuse, Dewey refused to accept the explicit separation between facts and values present in the attempts of the early critical theorists to drive a wedge between technoscience as concerned simply with the gathering of facts and their deployment by means of instrumental rationality, on one side, and a realm of meanings and values developed in the context of a lifeworld (a lifeworld that for Habermas came to include both communicative action and emancipatory action) on the other. For Dewey, as for Feenberg, technological decision making is at each fork in the road precisely about which of many possible values will be secured.

For Dewey, moreover, decision making in the spheres of technoscience is never a matter of starting from scratch. It operates against the backdrop of two sorts of assumptions: the first set includes those things which are in fact valued, and the second set includes those things that experimental deliberation has proven valuable. For Dewey, there are no pure, contextless facts ready to be gathered by activities of putatively value-free technosciences. Instead, it was Dewey's view that the technosciences operate in much the same way as do other areas of human inquiry: facts are always facts-of-a-case, selected by individual human agents or groups of them, embodied at a particular time and place and carrying forward a particular history against a particular cultural backdrop. For Dewey and Feenberg, but not for the early critical theorists, there is no contextless technoscience.

Perhaps even more important, however, both Dewey and Feenberg honor the idea that means and ends are not isolated from one another, that is, that in productive forms of technology neither means nor ends should be viewed as dominating the other. Feenberg's position is clear enough in his discussion of the ways in which various theories of technology have tended to treat this issue.

> Deterministic theories, such as traditional Marxism, minimize our power to control technical development, but consider technical means to be neutral insofar as they merely fulfill natural needs. Substantivism shares determinist skepticism regarding human agency but denies the neutrality thesis. Ellul, for example, considers ends to be so implicated in the technical means employed to realize them that it makes no sense to distinguish means from ends. Critical theories, such as Marcuse and Foucault's left dystopianism, affirm human agency while rejecting the neutrality of technology. Means and ends are linked in systems subject to our ultimate control. This is the position defended here, although I work it out differently from Marcuse and Foucault.[17]

Feenberg's difference with Foucault and Marcuse on this issue, as well as with what he calls the "common sense" view, seems to be as follows. For the "common sense" view, technology is neutral and is thus available to serve values and ends that are formulated independently. In the views of Foucault and Marcuse, however, "choices are not at the level of a particular means but at the level of a whole means-ends systems."[18]

Feenberg's own view is similar to that of Dewey. He posits two dimensions of what he terms "technical objects." The first is their social meaning, and the second is their cultural horizon. The point of the first dimension is that engineering goals hardly ever have the last word, even when successfully articulated. Although it may turn out to the disappointment of the engineers in question, social meanings, some of which are quite different from original engineering goals, also enter into the life of technical and technological objects. On this functionalist approach, straight-line instrumentalism gives way to the ramification of multiple possibilities. As Feenberg puts the matter, "[D]ifferences in the way social groups interpret and use technical objects are not merely extrinsic but make a difference in the nature of the objects themselves."[19] Dewey would, of course, have applauded this conclusion.

The second hermeneutic dimension, the cultural horizon, refers to cultural background assumptions. In the medieval period of the Latin West this involved a preoccupation with religious signs and symbols, and in our own milieu it involves "rationalization." Apparently unaware that he is echoing remarks that Dewey made more than six decades ago, Feenberg has concluded that "technology is thus not merely a means to an end; technical design standards define major portions of the social environment."[20]

For Feenberg technological choices are made by "social alliances." Such alliances appear to be more or less what Dewey termed "publics" in his 1927 book *The Public and its Problems*. Here is Feenberg: "A wide variety of social groups count as actors in technical development. Businessmen, technicians, customers, politicians, bureaucrats are all involved to one degree or another. They meet in the design process where they wield their influence by proffering or withholding resources, assigning purposes to new devices, fitting them into prevailing technical arrangements to their own benefit, imposing new uses on existing technical means, and so on. The interests and worldview of the actors are expressed in the technologies they participate in designing."[21] Feenberg's invigorated constructivism holds that "technology is social in much the same way as are institutions."[22] In Feenberg's vision of a "deep" democratization of technology, his alternative to technocracy, the activities of such social alliances will be wedded to "electoral controls" on technical institutions.

As I have already indicated, this vision, and the detailed analysis that supports it, of increasing electoral control by overlapping networks of educated and informed publics over the various "social alliances" that contribute to technoscientific decision making, is precisely what Dewey was arguing for in 1927 in *The Public and its Problems*.

In that work Dewey was highly critical of technological determinism. "There are those who lay the blame for all the evils of our lives on steam, electricity and machinery. It is always convenient to have a devil as well as a savior to bear the responsibilities of humanity. In reality, the trouble springs rather from the ideas and absence of ideas in connection with which technological factors operate."[23] Further, "the instrumentality becomes a master and works fatally as if possessed of a will of its own—not because it has a will but because man has not."[24]

Nor is this what Feenberg terms the "common sense view" of technoscientific neutrality. For Dewey, technoscientific artifacts teem with meanings, and this because such artifacts are the subject of intent and desire, and intent and desire are inevitably social in nature. "Primarily," writes Dewey in 1925, "meaning is intent and intent is not personal in a private and exclusive sense. . . . Secondarily, meaning is the acquisition of significance by things in their status in making possible and fulfilling shared cooperation."[25]

In 1939 Dewey specifically rejects the idea, still held by the critical theorists who were now working a stone's throw from his office at Columbia University, that "science is completely neutral and indifferent as to the ends and values which move men to act: that at most it only provides more efficient means for realization of ends that are and must be due to wants and desires completely independent of science."[26] In other words, he rejects the split between a world of technoscientific facts and a lifeworld of meanings and values.

In 1946, in a revised introduction to *The Public and its Problems*, Dewey puts this even more clearly.

> Science, being a human construction, is as much subject to human use as any other technological development. But, unfortunately, "use" includes misuse and abuse. Holding science to be an entity by itself, as is done in most of the current distinctions between science as "pure" and "applied," and then blaming it for social evils, like those of economic maladjustment and destruction in war, with a view to subordinating it to moral ideals, is of no positive benefit. On the contrary, it distracts us from using our knowledge and our most competent methods of observation in the performance of the work they are able to do. This work is the promotion of effective foresight of the consequences of social policies and institutional arrangements.[27]

How is this "promotion of effective foresight of the consequences of social policies and institutional arrangements" to be effected? Dewey cannot tell us directly, for his instrumentalism incorporates perspectivism, contextualism, and fallibilism. But if he cannot do this, he can at least discuss the conditions under which such a "great community" will be possible. Such conditions include the free flow of information that is secured by means of experimental inquiry, among various overlapping publics which refine and express their interests in ways that make them amenable to discussion and compromise, and an educational system that is committed to the development among children of an intelligence of the type that prepares them for participation in a great community. Experts will be relied on not to make policy but to clarify for the various publics the various consequences of alternative scenarios. It will require that the tools and methods of technology be employed to assure a level of material and emotional security that is the precondition for such a community.

> If the technological age can provide mankind with a firm and general basis of material security, it will be absorbed in a humane age. It will take its place as an instrumentality of shared and communicated experience. But without passage through a machine age, mankind's hold upon what is needful as the precondition of a free, flexible and many-colored life is so precarious and inequitable that competitive scramble for acquisition and frenzied use of the results of acquisition for purposes of excitation and display will be perpetuated.[28]

These remarks anticipate Feenberg's claim that technology will be democratized not solely, or even primarily, through the legal system, but through greater "initiative and participation" that would result in the "creation of a new public sphere embracing the technical background of social life, and a new style of rationalization that internalizes unaccounted costs borne by 'nature'."[29]

In sum, Feenberg follows Dewey on the following points. He has (1) moved from an essentialist to a functionalist understanding of technology, (2) developed a vigorous form of social constructivism, (3) rejected a Heideggerian-type romanticism in favor of a naturalized technology, (4) rejected the critical theorists' notion of technology as ideology, (5) accepted the idea that the project of Enlightenment rationality is not as much of a threat as the critical theorists had imagined, (6) proposed the idea that technical decisions are made within a network of competing factors in which one weighs various desired ends against one another, (7) warned against the reification of the results of inquiry as if they had existed prior to inquiry (Dewey's "philosophic fallacy"), and (8) recast technology in a way that crosses the line between artifacts and social relations.[30]

Did Dewey go beyond Feenberg? The short answer is yes. Dewey developed a philosophy of education and a deep analysis of "actor networks," which he termed "publics." He also developed a detailed philosophy of democracy, which is the centerpiece of his philosophy of technology. Taken with his theory of inquiry, and taken seriously, these aspects of Dewey's work provide the context for changing the way we talk about technology. My point in this chapter, however, has been to suggest that Feenberg's progress toward a pragmatic reading of the philosophy of technology is the right move at the right time.

Notes

1. See Joel Anderson, "The 'Third Generation' of the Frankfurt School," *Intellectual History Newsletter* 22 (2000) and available at <http://artsci.wustl.edu/~anderson/criticaltheory/3rdGeneration.htm>.
2. Andrew Feenberg, *Questioning Technology* (London: Routledge, 1999), 136. Feenberg devotes about a half page to Dewey, but dismisses his view as exhibiting a "rather uncritical confidence in science and technology."
3. Ibid., 147.
4. Ibid., vii.
5. Ibid.
6. For Horkheimer and Adorno, since they identify technology with instrumental rationality, there is an unbridgeable gulf between technology and the values of the human sciences. The same is true for Heidegger and for Habermas.

Marcuse's version of this position was considerably more flexible. He thought that scientific technology might be reformed under the proper conditions. A necessary condition for such reform would be the reform of political life.

7. Ibid., viii.
8. Ibid., x.
9. Anderson.
10. Feenberg, *Questioning*, xiii, 9. See also Andrew Feenberg, "Subversive Rationalization: Technology, Power, and Democracy," in *Technology and the Politics of Knowledge*, ed. Andrew Feenberg and Alastar Hannay (Bloomington: Indiana University Press, 1995), 4.
11. Feenberg, *Questioning*, viii, quoting Ihde, 1990:128.
12. See Melvin Kranzberg, "The Information Age," *Computers in the Human Context: Information Technology, Productivity and People*," ed. Tom Forester (Cambridge, MA: MIT Press, 1989), 30.
13. Feenberg, *Questioning*, viii.
14. Ibid., 210.
15. Ibid., viii.
16. Ibid., 213.
17. Ibid., 9.
18. Ibid., 7.
19. Ibid., 10.
20. Ibid., 97.
21. Ibid., 10–11.
22. Ibid., 11.
23. John Dewey, *The Later Works*, 1925–1953, ed. Jo Ann Boydston, 3 volumes (Carbondale: Southern Illinois University Press, 1981), v. 2, 323.
24. Ibid., v. 2, 244.
25. Ibid., v.1,142.
26. Ibid., v. 1, 160.
27. Ibid., v. 2, 381.
28. Ibid., v. 2.371.
29. Feenberg, "Rationalization," 19.
30. Feenberg, *Questioning*, 201 ff.

PART 2

The Politics of Technological Transformation

CHAPTER SIX

GERALD DOPPELT

Democracy and Technology

Democratizing Technology

Andrew Feenberg takes up the important task of developing a nonessentialist philosophy of technology. His aim, in a series of books written over the last decade, is a more democratic politics of technical decision making and a more rational design of our built environment.[1] To this end, Feenberg challenges the dominant modes of experiencing and understanding technology. He argues that the dominant essentialist model(s) of technology imprisons us in a world made by experts who use claims of expertise to exclude the voices and vital human interests of those lay groups most affected by it.[2]

On the essentialist image, technology and the built environment are perceived as determined, by necessary imperatives of efficiency and special bodies of expert professionals who enjoy a monopoly over knowledge of these imperatives.[3] The development of technology is seen to obey an autonomous and value-neutral logic in which science-based, technical elites (engineers, city planners, physicians, architects, etc.) realize ever more effective and reliable means to attain the necessary, incontrovertible goals of modern society. As such, existing technology at any particular moment in time appears to have a self-evident rationality that repels the possibility of authentic ethical choice and political debate. Of course such debate may arise over the proper use of technology—including questions of access and distribution. But the technology itself is seen as essentially outside all such political perturbations, marking the necessary framework of all rational action.[4]

Feenberg offers a theoretically powerful, empirically well-documented critique of this dominant understanding of technology. On his analysis, it masks the particularity, historicity, contingency, interest-ladenness, and politics of every specific technology that we confront in our built environment: cities, buildings, hospitals, clinical trials, machines and devices of all sorts, factories, etc. Each such technology embodies a design, and an underlying technical code that embodies established experts' determination of what is and is not a relevant factor in designing this or that sort of thing. In turn, Feenberg shows that every such design and technical code embodies particular peoples' decisions/power over which among many possible considerations, interests, values, costs, functions, and voices are to be included and which excluded in that technology.[5]

Feenberg thus teaches us to ask certain specific questions about any given technology. He establishes the social reality of these questions by illuminating contemporary cases in which lay groups have asked these questions and transformed technology in the process: "Who determined how this thing would be made, with whose or what purposes in view and out of view, at whose expense, in the context of what relations of power, and through what institutional or social process?" When lay actors or users of technology raise such questions, the paralyzing experience of an implacable technology or environment is dissolved and human agency is restored. By this route, Feenberg is able to explain and justify several real-life cases: how AIDS activists were able to transform clinical trials and experimental therapies from researchers' pure devices of medical knowledge to technologies of care and personal autonomy for people with incurable, fatal illness;[6] how movements of people with disabilities were able to transform the design of streets and buildings to incorporate ramps and thus their access and agency in relation to public life;[7] how millions of users of the French Teletel system were able to transform it from an information to a communication technology.[8]

On Feenberg's nonessentialist perspective, technology emerges as the embodiment of a social process in which empowered groups of experts choose to express certain sets of specific interests and standards in specific technologies, which in turn are reexperienced, challenged, and redefined by their users. The users bring their own meanings and interests to bear on the world, which can be different than those of its designers. Feenberg's work thus develops a new conception of technology as the site of contingent political contestations between different groups of actors over the interests, purposes, and meanings that will be invested and encoded in the built environment. He effectively achieves a demystification of technology, which parallels the demystification of authoritarian state power achieved by the classical liberal philosophers, and the demystification of capitalist political economy/market relations accomplished by Marx and his heirs.

Democratizing Technology and Society

Feenberg's most recent work seeks to go well beyond the demystification of technology by arguing that his nonessentialist philosophy paves the way for a democratization of technology, and indeed, a radical democratization of society itself.[9] Feenberg looks to his account of the politics of technology as the basis for unifying a whole disparate array of new social movements of reform. He sees in his philosophy of technology the basis for an overarching ideal of radical democracy and a utopian vision of an "alternative" or "redeemed" modernity.[10]

In the rest of this chapter, I argue that Feenberg's nonessentialist philosophy of technology lacks the ethical resources required by these grander aims and in any case misconstrues the aims themselves. My argument revolves around three points of criticism. First, I maintain that Feenberg's conception of democratization of technology is simply too abstract to serve as a standard of emancipation. Secondly, I argue that he needs a more substantive conception of democratic ideals—especially of democratic equality and rights. Without that, I argue, he cannot distinguish between challenges to, or changes in, technology that strengthen democratic ideals, from those that do not, or have a minimal democratic impact. Thirdly, I claim that Feenberg's theory of democratization of technology faces powerful normative obstacles which it does not take into account. In particular, nondemocratic technology, however it is interpreted, rests on our society's powerful Lockean moral code of private property, and not simply on the technocratic ideology of essentialism and value-neutral efficiency.

In sum, I argue that there are large theoretical, ethical, and political gaps between Feenberg's achievement (i.e., the demystification of technology) and his broader vision of democratization and alternative modernity. My point is not that Feenberg's nonessentialist philosophy is no aid to the broader vision, but rather, that the broader vision requires, in addition, a different kind of analysis and ethical critique of modern liberal democracy than what is now central in Feenberg's approach.

"Participant Interests"

Ultimately, Feenberg's nonessentialist conception of technology is supposed to show all lay actors who are impacted by technology not just that they can change it, but that they do or may have good reasons for doing so. But what provides the "good reasons" in Feenberg's model? The answer is given by his notion of "participant interests."[11] Lay actors who have to use or live with the environment built by others either have or may develop various "participant interests," which have the potential of being embodied in its design and structure. Intuitively, the notion is clear enough. "Participant interests" refers to the ways that the personal welfare of various people—participants, users, and third parties, etc.—is commonly impacted, for better or worse, by a technology, making them into a group bound together by such "participant interests," affirmed or frustrated by that technology, as the case may be.

This notion must bear considerable weight in Feenberg's overall argument against essentialism. After all, the essentialist who equates technology with efficiency and rationality sees that its design embodies persons' interests. The essentialist perceives these interests as typically uncontroversial, inescapable, and universal in modern life—shared by users, designers, and the public, alike. Owners, designers, workers, and buyers all share an interest in making a car that will sell, run, etc. The collective emergence of marginalized "participant interests" reveals the one-sidedness, contingency, and politics of technology; that is, particular users' interests and voices excluded from the design process and technology. In seeing what a technology excludes, what it is not but could be, we gain a clearer and truer grasp of what it is—the bias of technology beneath the guise of efficiency and rationality.

Feenberg's notion of participant interests must also bear considerable weight in his argument for democratization. For, it is certain participant interests that must provide (1) the motivation, (2) the justification or reasons, and possibly (3) part of the criterion, for the democratization of technology. The establishment of marginalized participant interests, or ones once marginalized but now included in the technical code, shows that technology could be other than what it is. But this falls well short of defending a conception of what technology ought to or should be. Which participant interests should be accommodated within a democratized technology, or alternative modernity? Which technologies or aspects of our built environment ought to be democratically transformed; or more importantly, in accordance with what standard of emancipation, or human well-being? This is the key ethical problem that requires exploration by a democratic critique of technology.

Feenberg may reply that this is not his problem or project. Having shown that technology can be changed, it is up to the users, the public, to assert their marginalized participant interests and to determine how the built environment ought to be changed. Once technology is demystified, a desirable democratization of technical design inevitably follows, so the argument might go. I see two large problems for this strategy, which in any case doesn't seem to be

Feenberg's. I will refer to these two problems as the "Which Interests" problem and the "Private Property" problem.

The Problem of "Which Interests?"

The first problem is this: it is clear that not every participant interest, or challenge to technology is legitimate, morally justified, or a victory for democratization. There are reactionary challenges to technology—ones that have led or would lead to a less rational, equitable, or democratic technology.[12] Without an ethical standard, how can we determine which trade-offs, whose interests, what challenges contribute to a more, or less, democratic rationalization of technology? There are also cases of successful challenges to technology that are neither reactionary nor ambiguous but don't really democratize the technical code. To my mind, Feenberg's example of Teletel falls into this category.[13] In general terms, it is a case where large numbers of customers put a technology to uses alien to its designers' will and design, which are then exploited to produce a more marketable and lucrative technology with these new uses now built into the design. While this may be desirable, such changes of technology in response to consumers' initiatives or preferences follow the logic of market rationalization, not democratization.

Feenberg presents some intuitively quite attractive contemporary examples of successful lay challenges to established technology. The struggle of disabled people for access ramps and that of AIDS activists for a different technical code governing the design of clinical trials are two such examples. I examine these cases below in order to determine whether and on what basis each can be characterized as democratizing technology. But the cases do not provide an explication and defense of a normative standard for judging which interests and challenges might produce a more rational, democratic, morally defensible technology, and which do not. This is what Feenberg's argument for democratizing technology and an "alternative modernity" requires but lacks.

Presumably, Feenberg does not want to embrace any and all participant interests or challenges that happen to surface or succeed. For one thing, he implicitly acknowledges that lay actors may have participant interests which they fail to interpret and articulate as claims that can gain wider social recognition and legitimacy.[14] For another, he implies that even when they do get articulated as claims or demands, not all such claims deserve recognition and legitimacy.[15] On the other hand, Feenberg cannot justifiably pick and choose the cases he finds attractive and ignore others, without some systematic ethical justification. The development of a standard of justification is not external to Feenberg's project—a gratuitous appendage he can take or leave. Without it, there is little reason to accept his claim that the emergence of new participant interests and lay challenges in technosystems informs a process of democratizing technology and creating a more rational "modernity." Rather, all his argument would establish, against essentialism, is that technology is not destiny; because local shifts in relations of power and interest alter technology, in some cases for the good of particular groups.

It is possible that my critique could be preempted by Feenberg's conception of democratization and his analysis of cases of it—either of which might contain the normative standard(s) that, I have argued, he requires. Before turning to examine these contributions, I turn to the second problem facing his critical, nonessentialist theory of technology—the "Private Property" problem.

The Problem of "Private Property"

Users' marginalized participant interests often fail to emerge or gain social recognition as legitimate claims for technical change because the established technical code is taken to embody the will, property rights, and legitimate interests of the owners of the technology. In the modern world, a great deal of technology is private property. Its designers act in the name of the owners, and their rights as owners, to determine the technical code in accordance with their economic interests. In these contexts users' participant interests and challenges quickly confront the claims of designers and experts that they alone have the right to decide, to determine technology. Feenberg's theory pictures these conventional prerogatives of expert authority as always an expression of technocratic ideology—the very essentialist image of efficiency he demystifies. But in the common case where technology is private property, the rights and authority of the designers/experts really rests on the fact that they are employees, representatives of capital. The rights of the designers to exercise authority rest not just on their expertise and the logic of efficiency, but on the rights of private property, and the Lockean moral code of ownership and free-market exchange. To this extent, Feenberg's critique of essentialist technocratic logic is insufficient to explain, motivate, or justify the democratization of technology. For, in modern society, it is the powerful moral code of private property, not just technocratic ideology, that opposes the translation of users' or workers' participant interests into legitimate rights to reshape technology. Do they have such rights? What rights? How are they ethically justified? Those key ethical issues cast a pall over Feenberg's democratic euphoria.

If I am right, in the common case where technology is private property, the technical code(s) is embedded in the moral code of Lockean ownership and the rationality of capitalist market relations. The experts who claim and exercise rights of exclusive control over technology are widely perceived not just as authoritative arbiters of efficiency, but as the designated agents of the will and rights of owners, and less directly, the will and rights of consumers who get to "vote" on technical design with their dollars. In these contexts, challenges to technology, based on participant interests, involve challenges to the rights of private property and modern society's powerful Lockean moral code. To draw on one of Feenberg's examples, workers (or their unions) may challenge production technology on the basis of their participant interests "in such things as health and safety on the job, educational qualifications and skill levels, and so on."[16] When they do so, their claims may be discredited or undermined not simply by experts' judgment concerning what is and isn't feasible, efficient, etc. but by owners' or top managements' "right" to reject such changes as unprofitable, unnecessary, or incompatible with company policy as they define it. Regardless of how the dominant technical code is perceived, if it embodies the will of the owners and company policy, then many see it as legitimate, and workers' or users' challenge to it as lacking moral force.

Because many technical codes are grounded in the Lockean moral code of private property, dominant technology is provided with powerful rights-based protections just as important as technocratic essentialism. Rational challenges to technology in these cases pushes beyond the logic of Feenberg's participant interests to the issue of users', workers', or participants' basic rights. If established technology can be reasonably reinterpreted as a violation of these actors' basic rights or entitlements, this may provide a good reason for transforming dominant understandings of the rights of private property, not simply the prerogatives of expert authority.

The private property problem forces a reorientation of Feenberg's project, moving it beyond the critique of essentialism and the discourse of participant interests. Rather, his argument needs to develop an account of the logic through which some participant interests but not others have been or can be reasonably represented as legitimate claims of right, counterbalancing rights of private property. Whatever else it may involve, a democratization of technology goes well beyond its demystification. It involves not just a reinterpretation of technology, but an ethically well-argued reinterpretation or revision of the Lockean and liberal-democratic moralities of right.[17] As I argue in the last section, the aim is to develop a general ethical standard or conception powerful enough to establish a link between some but not all challenges to technology (and rights of private property) and the substance of liberal-democratic ideals, properly interpreted.

Can such a standard be identified in Feenberg's conception of democratization or his examples of it?

Changing Technology and Democratizing Society

My aim in the remainder of this chapter is to argue that Feenberg's theoretical account of democratization and his examples of democratizing technology do not provide an adequate normative standard or set of standards. When we characterize society or social change as more or less democratic, we may operate with very different standards in mind concerning its institutions, practices, and ideals. It is useful to distinguish standards concerning democratic models of political agency from ones concerning democratic models of equality or individual freedom/rights.

We may focus on how power is exercised, who can or cannot, does or does not have a voice in the key decision-making practices of the society at various levels of social life. I'll call various standards with this focus ideals of democratic political agency, or political agency conceptions. On the other hand, we may focus on the impact of a society's decision-making practices on its structure of democratic equality or freedom: the degree to which its citizens and groups enjoy equal individual rights, freedom, opportunities, essential resources, and statuses in the society. Agency is involved here, but it is agency in the sense of individual autonomy—personal control over one's own life and activity—rather than political participation/agency in the above sense. I will call standards that focus on this second dimension ideals of democratic equality, or equality conceptions—though it is often personal agency which equal rights, opportunities, resources, etc., protect. Political theories disagree on how they interpret the substance of one or both of these two sorts of democratic ideals—political agency and equality—and which they see as more normatively important or fundamental to a democratic society.

For example, the liberal tradition from classical theorists to John Rawls sees democratic political agency, representative government, as having primarily an instrumental value. It is the only or best arrangement for protecting individual freedom, rights, equality of opportunity, and market freedoms. Other theorists (civic republicanism, radical democracy, etc.) embrace far more ambitious models of democratic political agency, which is seen as having intrinsic value or virtues, as, or more, central to "true democracy" than individual freedom or the equality that protects it.[18] Feenberg is not concerned with these normative disagreements.

Yet, it is clear from his treatment of theorists of democracy such as Barber and Sclove, that he is operating with a "political agency" conception or ideal of democracy.[19] With Barber and Sclove, Feenberg wants to reject the liberal ideal of democracy I characterize above. Following Barber's model of "strong" democracy, Feenberg accords central value to a populist, participatory politics involving local collective action, direct citizen intervention, and bold social movements, exemplified by AIDS activists, environmentalists, feminists, and other groups of lay actors. Following Sclove, Feenberg holds that the key problem for this participatory model of radical democratic political agency is how to apply it to technology. Indeed, his formulation of this problem and the way he seeks to resolve it constitute impressive insights.

Both standard liberal and radical participatory models of democratic agency define the relevant units of agency, the public, and representation by means of conventional geographic boundaries. Thus, on these various models, the relevant political actors are variously identified as the citizens of the nation, the citizens of this or that municipality, the employees of a hospital or factory, etc. But, as Feenberg argues, the lay public that might exercise democratic control over technology cannot be identified by such conventional political boundaries.[20] Modern technology implies the "fragmentation of technical publics"—a proliferation of diverse subgroups of users, each of whom bear different practical relations to a technosystem and none of whom necessarily occupy one and the same conventional geographic or political boundary.[21] Who then is supposed to exercise democratic political control over technology, and on what basis?

Feenberg's intriguing answer is that technical networks create *new* political subjects—e.g., ill people seeking access to experimental drugs or clinical trials, or users of a new technology such as the French Minitel.[22] While such subjects defy conventional geographical and political boundaries, their common practical relation to a technology may give them common "participant interests" which in turn can become the basis for democratic political agency concerning that technology's impact on them. In dialectical terms, a technology, and the experts that control it, create their own "other," specific groups that develop new common interests that prepare the way for a dialectical overcoming, and more harmonious "whole," of technical and human relations. Through such a dialectic, the users, participants, customers, third parties impacted by their common practical relation(s) to technology theory become political agents—precisely the agents required by a democratization of technology! A beautiful theory, indeed.

But, at this point, my earlier critique re-arises. The very fragmentation of technical publics stressed by Feenberg's incisive analysis of modern technology seems to bring in its wake a fragmentation of participant interests, agency, and the very meaning or prospect of democratization. What standard of democratization is operative in his argument? Is he willing to see democratization whenever any politically marginalized lay group(s) of users exercises power over technology? Or does it also depend on the substance of their interests and demands? Doesn't it depend on the relationship of what they do to the broader democratic ideals of political agency and equality? Doesn't it depend on how the group justifies its interests, how other groups' interests and political agency are impacted, and how other groups interpret what has been accomplished?

The fact that one subgroup of users of technology gains some power over it should not necessarily count as democratization, especially if the change comes at the price of disempowering or excluding other broader groups of users with basic rights, opportunities, or interests at stake. Suppose a group of industrial or office workers succeed in gaining a marginally safer, or more pleasurable, or easier to operate, technology in the workplace. Imagine that the "cost" is that others who are not consulted lose their jobs or skills. Or, that thousands or millions of consumers lose access to a basic good or service as a result. Worse, imagine that those who benefit and those who are disadvantaged or harmed by the change in technology are already divided by differences of race, class, ethnicity, or gender. Without taking such zero-sum conflicts of group interest, power, and identity into account, we cannot reasonably evaluate whether or not the change in technology should be, or in fact will be, seen as a democratization of technology. And, we require a defensible ethical standard of democratization to ground such evaluations, and to transform the politics of technology.

Democratic Equality

If I am right, Feenberg's vision of democratizing technology and an alternative modernity must be grounded in an ethical account of the interests that might define, justify, and motivate this emancipatory project. What I've called the Lockean moral code does not just provide rights-based protections of technology as private property. As a modern ethos, it shapes users' very desires and interests in ways that tend to bind them to the rationality of established market and technical relations.

Feenberg's work is permeated by an ironic tension or contradiction. His theory of democratization rests exclusively on the "political agency" paradigm. But his examples of successful democratization, as well as his evaluations of absent, failed, or flawed democratization, implicitly rest on the "equality" paradigm: a view of the sorts of substantive interests, the specific changes in the technical codes at issue in a genuine democratization of technology. For Feenberg, the obstacle to democratizing technology is not primarily political powerlessness (the political agency paradigm), but users'/participants' failure to recognize or assert the "right" interests and changes in technology (the equality paradigm).

When we turn to Feenberg's accounts of successful democratizations of technology the same points emerge, with some clues concerning how to address them. He characterizes the movement of disabled people for barrier-free design or ramps on streets, public buildings, businesses, etc., as the most "compelling" example of a democratic politics of technology.[23] The dominant technical code for designing sidewalks could "rationally" or "efficiently" exclude ramps, as long as disabilities count as purely personal problems, irrelevant to the design of public spaces. But disabled people comprise a large population with a powerful participant interest in "mainstream social participation."[24] Once they mobilize and gain public recognition of this interest, many technical features of the built environment are transformed. Why is this such a compelling case of democratizing technology? We need to attend to the paradigm of democratic equality and ask who is this group of "users," where do they stand in society, what have they been denied, and what is the ethical significance of the technical change they demand?

The movement for barrier-free technology was identified by disabled people, and eventually, most of the public, as a struggle for basic civil liberties, citizenship, the right to enjoy the same access to public buildings, social life, urban mobility, as other Americans. Thus, disabled peoples' transformations of technology is democratic first and foremost, because it involves a victimized group's gaining (1) the same ability to exercise fundamental rights and liberties as other citizens, and (2) public recognition that they can do so, and deserve to. In sum, the action of the disabled is about the ideal of democratic equality. It counts as a democratization of technology precisely for that reason.

In the next section, I examine Feenberg's treatment of AIDS activists and seek to determine the criteria of democratization that we should employ to evaluate it.

Democratizing Experimental Medicine

Feenberg provides his most explicit, detailed normative argumentation concerning participant interests in his account of the challenge of AIDS activists to the technical code underlying clinical trials of experimental drugs.[25] On the technical code of medical experts, there is a sharp distinction between research and treatment or care. Clinical trials involving the testing of experimental drugs fall under the category of scientific research aimed at gaining knowledge of the drug's medical value. Clinical trials are not a form of legitimate treatment or care because the medical efficacy of the drug has not yet been scientifically established.

This technical code shapes the design and regulatory policy governing clinical trials and access to experimental drugs more generally. Experimental drugs should be used exclusively as objects of research, not modes of treatment, until their medical value is scientifically established and certified. This provides the rational basis for the design of clinical trials. Because they constitute research, not treatment, it is entirely efficient and appropriate to limit patient access to clinical trials, use placebos to determine the causal efficacy of the drug, and employ elaborate conventions of informal consent to protect participants from deception and false hopes.

In the story that Feenberg recounts, AIDS patients and activists demanded a radically liberalized and expanded access to experimental drugs and clinical trials. These are people with terminal and incurable illnesses. They have the largest personal stake in finding a cure or beneficial treatment. They want to be in a position to try unproven drugs that might help or work and to be participants in the life and death race for a cure, rather than doing nothing. So they demand changes in the policies, availability, and design of clinical trials, in order to gain access to experimental drugs and the research process.

Feenberg provides a moral justification for the AIDS activists' challenge to established medical technology based on its own internal moral code of providing care to the ill and suffering. He argues that upon analysis, clinical trials and access to experimental drugs are forms of medical treatment or care and not exclusively contexts of pure scientific research.[26] Modern medicine embraces "care" as one of its major professional callings, even if it has been somewhat eclipsed by the technical pursuit of "cures." On Feenberg's illuminating interpretation, what the AIDS activists accomplished is the leveraging of modern medicine's moral code to revise its technical code. As a result, previously well-established policies surrounding drug trials (e.g., limiting participation to statistical minimums, the employment of placebos, excluding subjects with prior experience in drug trials, etc.) could be rationally modified in some cases to accommodate subjects' now "legitimate" interest in participation.

Does Feenberg's argument concerning this powerful example establish that it is a case of democratizing technology? Does it provide implicit clues concerning the normative standard(s) of democratization we should employ to evaluate and generalize such cases of technical change? It would seem that the change in the technical code defended by Feenberg amounts to the important but essentially modest claim that "a good medicine," when and where it is feasible, will "design experiments that serve patients while simultaneously serving science."[27] But even this claim goes well beyond what his reading of AIDS activists establishes. This case shows that the interest in experiment as care is an urgent, legitimate factor in design for persons with incurable terminal illness; it doesn't establish the same point for any and all persons with any illness whatsoever, which is what some of Feenberg's formulations suggest. Nothing about the case of people with AIDS speaks to the status of people who desire to enter experiments that test new drugs for skin rash and their interest in the care such participation may bring.

Feenberg poses the key normative question inspired by his nonessential theory of technology: Does the desire and demand of AIDS victims for access to drug trials deserve public recognition as a legitimate participant interest and a rational basis for revising the technical code governing the design of experimental medicine? The way Feenberg poses and answers this question fits well with the general critique I have made of his argument for democratizing technology. How so? The AIDS activists achieved some success in expanding access and revising the technical code of clinical trials as pure scientific research. But, as far as I can tell, Feenberg is not content to take the mere fact of this marginalized group's empowerment as "the" mark of a democratic rationalization (improvement) of experimental medicine. Rather, he seeks to provide a moral justification for taking AIDS victims' desire/demand for access as a good ethical reason for revising the technical code of clinical trials. But what he says in this case may provide clues for developing a general theory.

In order to evaluate the impact of any particular case of technical politics on the democratization of technology and society, we need to ask *who* is this group of users challenging technology, *where* do they stand in society, *what* have they been denied, and *what* is the ethical significance of the technical change they seek, for democratic ideals? These questions call the attention to features of the AIDS example that are relevant to the standard(s) of democratization but are ignored in Feenberg's treatment.

When we ask whom the AIDS activists represent, it is clear that they are *not* simply a group of citizens with a deadly or incurable illness. It is a stigmatizing illness culturally associated with a group oppressed by homophobic fear, hatred, prejudice, and discrimination. The struggle of AIDS activists for access to treatment, care, agency, choice, etc. challenges the moral code that sustained homophobia and sanctioned the early record of sparse funding for AIDS research and care. Their success achieves a significant measure of human dignity, agency, respect, and citizenship—even though they do not win a right to care, or a right to participate in clinical trials, as Feenberg admits. It is this positive impact on the ideal of democratic equality that makes it a powerful case of democratizing technology. As in the case of the struggle of the disabled for ramps, it is not exclusively or primarily the exercise of political power by lay actors that provides the criterion of democratization. Rather, we need to also look at the impact of these struggles over technology both on the moral and political standing of oppressed groups and the long march to genuine democratic equality for all.

Toward an Ethics of Democratic Equality

Let us return briefly to the example of the struggle of the disabled for ramps, in order to summarize my argument in this chapter. This example highlights the strengths and weaknesses of Feenberg's theory of technology and the path of future research on its democratization.

The movement of disabled people for barrier-free access requires both a demystification of technology and an ethical reconstruction of the ideal of democratic equality and the meaning of citizenship/personhood for the disabled. Following Feenberg, for democratic change to occur, lay actors/users need to see that existing sidewalks, bridges, building entrances, restrooms, classrooms, etc. (without entry access for the disabled) exclude the interests and voices of some group(s) of users. Feenberg's theory accomplishes this important task, important because it constitutes a necessary condition for democratization. But this much does not necessarily provide public or private parties with good reasons to support a design change or force an opening up of decision making, both of which always involve costs of some sort to somebody (e.g., taxpayers). Where powerful moral rights, entitlements, and ideals at stake, actors must provide good reasons to abridge or override established notions.

As I have argued, Feenberg's theory does not possess the resources to accomplish this task. He doesn't develop any framework of ethical concepts for analyzing democratic ideals or showing how they enter into the cases of users' challenges to technology that he examines. His theory does not articulate any general ethical standard(s) of democratization that would enable us to evaluate particular challenges to technology or to define a liberating politics of technology, an alternative modernity. More generally, the influential discourses of anti-essentialism and social constructivism create a political vacuum, which can only be filled by affirmative ethical argument concerning the right and the good. The political philosopher who aims at a general democratization of technology confronts a multidimensional project—criticizing and revising, in tandem, dominant ethical understandings of technology, private property, democratic equality, and the morality of individual rights. The critical theory of technology cannot complete its mission without turning into a more thoroughgoing critical philosophy of liberal-democratic values. An Alternative Modernity is a modernity with a different vision of human relationships and the good life. The democratization of technology awaits the development of an alternative ethical understanding of the ends of modern society.

Notes

An earlier version of this chapter was read at the Pacific Division, American Philosophical Association meetings (April 3–5, 2000, Albuquerque, New Mexico) in an "Author Meets Critics" session on Andrew Feenberg's *Questioning Technology*. I am happy to acknowledge my debt to all of Feenberg's work and his valuable responses to my work on this occasion and many others. In writing this paper, I have also gained illumination from Robert Pippin's work on the philosophy of technology in "On the Notion of Technology as Ideology: Prospects" in *Technology and the Politics of Knowledge*, ed. A. Feenberg and A. Hannay (Bloomington and Indianapolis: Indiana University Press, 1995).

1. In this chapter, I will mainly focus upon theoretical formulations in A. Feenberg's, *Questioning Technology* (London and New York: Rutledge, 1999), which I hereafter abbreviate as "*QT*," *and* the historical cases in his *Alternative Modernity: The Technical Turn in Philosophy and Social Theory* (Los Angeles: University of California Press, 1995), which hereafter I abbreviate as "AM." Also see *Critical Theory of Technology* (New York: Oxford University Press, 1991). All three works make important contributions to Feenberg's philosophy of technology.
2. *QT*, vii–xvi.
3. Ibid., vii–x (14–17, 201–16).
4. Ibid., 16–17.
5. Ibid., 87–89 (142–43).
6. AM, 96–123 and *QT*, 127–28 (141–42).
7. *QT*, 141.
8. Op. cit., 144–69, *QT*, 126–28.
9. *QT*, vii–xvi (104–109, 131–47).
10. Ibid. 222–25.
11. Ibid., 140–42, AM, 104–109.
12. Feenberg acknowledges such cases and the problems they raise. See *QT*, 92–94.
13. AM, 144–69, *QT*, 125–28.
14. *QT*, 140–42.
15. Ibid.
16. *QT*, 140.
17. For my own attempts to advance such an argument see "Rawls' System of Justice: A Critique from the Left," *NOUS* 15, no. 3 (1981); "Conflicting Paradigms of Human Freedom and the Problem of Justification," *Inquiry* (Spring 1984); "Rawls' Kantian Ideal and the Viability of Modern Liberalism," *Inquiry* 31 (1988): 413–19; and "Is Rawls' Kantian Liberalism Coherent and Defensible?" in *Ethics* 99 (July 1989), 815–51.
18. For a recent and engaging contribution to this tradition, see I. Shapiro, *Democratic Justice* (New Haven: Yale University Press, 1999), and my critical study of it in *Inquiry* (Summer 2001) in "Does More Democracy Yield Greater Justice."
19. *QT*, 132–37.
20. Ibid., 134–47.
21. Ibid.
22. Ibid.

23. Ibid., 141.
24. Ibid.
25. Ibid., 96–120.
26. Ibid., 110–18.
27. *QT*, 141; AM, 109.

CHAPTER SEVEN

ALBERT BORGMANN

Feenberg and the Reform of Technology

Reform is often thought to be the touchstone of the significance a theory of technology can claim.[1] More precisely, the putative test is whether real reforms follow from a particular theory. Larry Hickman's recent book illustrates the phenomenon. It ends with a chapter titled "The Next Technological Revolution."[2] In it Hickman examines the major, primarily American, critiques and theories of technology to see what kinds of reform they offer or imply—kinds of *reform, nota bene*, and not of revolution. The Marxist confidence that the dialectic of history will produce a revolution come what otherwise may, that the most determined opposition will merely retard it, and the strongest support will only hasten it a little, that confidence has evaporated.

Hence, philosophers of technology have at least this in common when it comes to reform: it will not be easy or obvious. This understanding too is exemplified in Hickman's last chapter. He finds all reform proposals wanting because of their optimism, pessimism, thinness, or small scale. In other words, the suggestions, according to Hickman, overestimate or underestimate the feasibility of reform, they are restricted to just one thin layer of the technological culture, or they fail to take on the large structures of power. To this list of failings I would add the restriction to purely procedural reforms.

The criteria for real reform, implicit in these criteria, are political realism, cultural depth, structural comprehensiveness, and substantive content. They are more formidable than they look. Still, the first one seems eminently reasonable. Hickman rightly points out that we must start with the conditions we actually face—*hic Rhodus, hic salta,* as Hegel put it. But what could possibly be the measure of feasibility or possibility? One obvious answer is that the actual is also the possible. In practice this comes to the implicit expectation that feasible reforms have to look much like the reforms that have actually succeeded. There have been three such reforms in the second half of the twentieth century—civil rights, environmentalism, and feminism.

Most philosophers of technology in this country live left of center; these three reforms have their sympathy and have tuned their sense of change. Hence, any proposal that fails to resonate with this tuning simply does not register. The problem with this conception of reform is that for all its benign complexion it often takes technology for granted. To urge a socially more just distribution or an environmentally kinder production of goods and services is likely to leave technology itself unreformed.

While the advocates of the first criterion fail to grasp what a genuine theory of technology needs to do, the proponents of the second and third criteria demand too much of a theory of technology, and when presented with one that claims to meet their criteria, they tend to dismiss it as essentialist. A social theory that is both trenchant and sweeping is hard to find. Sweeping generalities are a dime a dozen. Likewise, some particularly arresting and penetrating insight or other can be found in most books on technology. A celebrated example are the low bridges on the Wantagh Parkway in New York State that kept the buses of people without wheels off the parkway. But this example does not generalize well. Social discrimination become stone can be found, but it is rare.[3] Usually it is the price of real estate, the expense of a meal at a fine restaurant, or the vigilance of doormen that keeps the poor from the rich.

Though breadth and depth are both dimensions of one phenomenon, viz., of the technological culture, plumbing depth is the more difficult requirement. Going for critical depth is much like insisting on cultural excellence, and when it comes to excellence, philosophers and social scientists are typically silent. The chief cause of that reticence is the rise of liberal individualism, shared by deontological and utilitarian ethics alike. Absent a substantive notion of the good life, the move instinctively taken by most social theorists is a retreat to procedural proposals, to calls, for example, for participatory democracy or for decentralization. But just as injustice is compatible with due process, procedural norms can be met by a sullen and shallow society.

The predicament of reform is made worse by the interrelation of the criteria of reform. We cannot tell how feasible a reform proposal is until we know its extent and incisiveness. How incisive a proposal can claim to be depends on its substantive norms. But we cannot decide whether a substantive proposal deserves to be taken seriously until we have determined its feasibility.

Thus, there seems to be no real chance of satisfying one criterion of reform without meeting all criteria.

Andrew Feenberg's recent work breaches this vicious circle. It occupies a uniquely important position in contemporary philosophy for a number of reasons. Feenberg, to start with, is currently the most prominent and productive philosopher in the area of technology and politics. He is an extraordinarily well read and cosmopolitan thinker. He has made original contributions to technology studies and political theory. He has fashioned a viable and hopeful philosophy of the Left. And along with Douglas Kellner, Feenberg has kept the heritage of the Frankfurt School alive and has examined and further developed its distinctive insights into the connections between culture, technology, and social justice.

Feenberg's most original and hopeful contribution is presented in his *Questioning Technology* of 1999. He addresses the problem of feasibility in Part Two of the book by arguing, against technological determinism, that technology in the narrow technical sense fails to have a unified character and fails to proceed in a unilinear direction. Especially in advancing the first of these two criticisms, Feenberg appeals to recent work in technological constructivism though he criticizes constructivism for remaining at the level of politically inconclusive episodes of design. The crucial issue is the political indeterminacy of technology. Here is a space for genuine reform.

An important feature of this space is the nature of its boundaries. They are drawn as much by technology itself as by politics. Technology, however specified, sets boundaries for human action and therefore shapes contemporary life much as legislation does. Feenberg shows that calls for political participation that overlook the "legislation" through technology can hardly be successful. In particular, the orientation of standard political theory sees power as organized in spatial terms—municipalities, countries, states, the nation. But information technology, one might say, organizes power and alliances in topological as well as metric space.

The standard way that the impact of technology on society is seen, Feenberg calls primary instrumentalization. It consists in the design and purpose that a device or a system is given by its sponsors and designers. Primary instrumentalization typically abstracts from context, aesthetics, and the involvement of users. Secondary instrumentalization designates the ways in which technological devices are appropriated by groups of individuals, that is, integrated into contexts, endowed with aesthetic and ethical significance, engaged by human beings, and adapted to their intentions within the "margin of maneuver" that inevitably is left by technology. Among his examples are the appropriation of Minitel in France for social purposes and the appropriation of the Internet by AIDS activists. Feenberg connects his idea of secondary instrumentalization with Simondon's notion of concretization and concludes by setting his politics of "a better way of life, a viable ideal of abundance, and a free and independent human type" between and beyond the socialist and capitalist visions.

Feenberg makes an incidental but important contribution to political theory, first by reminding us that "[p]olitical theory has yet to come to terms" with the politics that is powerfully and consequentially built into technology (131), and second by showing both through examples and structural analysis just where and how a liberating sort of politics has been launched within technology.[4] Whatever the devastations that have been inflicted on this country in the name of technology, few if any of Feenberg's readers would want to live elsewhere. All of us are grateful that we have access to governmental information, can call governmental agencies to account, can use the Internet to organize and promote our causes, can demand and criticize official justifications for environmental decisions, can change lax drunk-driving laws, can campaign against smoking, and much else. Too often we are unappreciative as citizens and unobservant as scholars when it comes to such democratic technologies. Feenberg generally does not chide us for being ungrateful; rather, he opens our eyes to a distinctive kind of technology, to its logic, to its antithesis, its concealments, and its occurrences. "We need a method," he says, " that can appreciate these occasions, even if they are few and far between, even if we cannot predict their ultimate success" (180). Such are the consolations of philosophy. Clear-eyed reflection lets us see what is in fact the case and where our best hopes lie. In replying to a review of *Questioning*, Feenberg explicitly rejects a global and radical response to technology.[5]

Measuring Feenberg's democratic technology against the criteria of reform I began with, we can see that it meets the criterion of feasibility, in part because democratic technology has been actually realized, and the actual was evidently feasible. Some of these technologies meet the standard expectations of social and environmental reforms, and such reforms neither question nor transform the underlying technology itself. It is one of the virtues of Feenberg's model, however, that it shows how austere and ambiguous a phenomenon a technology is in the strictly technical sense.

What is the Internet in and of itself as a technical object? For the Department of Defense it was a robust and resilient communications network. For Lawrence Lessig it is a cultural commons.[6] For AIDS activists it is a common bond and a tool of political action. Some of the Internet's technical features, to be sure, have politics built into them. Packet switching provides for the robustness and resilience the Department of Defense requires. The relative simplicity of the network keeps sophistication at the ends (the sender and receiver) of the communications channel (the "end-to-end" principle) and thus prevents biases in favor of certain messages and gives everyone, once admitted to the network, equal access.[7]

Is it possible to extend the feasibility of Feenberg's model down to this level of nuts and bolts? As it turns out, the fertility of the model increases its feasibility. We can subsume Jesse Tatum's work on the home power movement and Gordon Brittan's on wind turbines under the democratic technology of secondary instrumentalization.[8] The practitioners of home power go off the grid and construct from components that are available at different levels of adaptability an electricity generation facility and a household that is efficient in energy consumption. Similarly, Brittanic wind turbines with their beautiful sails and accessible machinery take electricity generation from giant and forbidding structures to an aesthetically pleasing human scale and out of the hands of highly centralized and bureaucratic organizations into the competence of fairly ordinary people. Granted, wind turbines have failed the feasibility criterion in a straightforward sense—there are very few of them. But the obstacles they have run into are not primarily technical or cultural. They are barriers of energy policy. Environmental pressure and scarcity of resources may yet lead to the realization of Brittan's project.

This extension of Feenberg's ideas also answers the requirement of cultural depth. Surely the home power and, if the project comes to fruition, the wind turbine people appropriate their world in a way that is better informed about energy policy and technology, requires and rewards more hands-on competence, and appreciates the weather, the seasons, and the climate more acutely.

As for the sweep and scope of Feenberg's theory, it teaches us to abandon all-encompassing ambitions. Technology is so extensive and complex a phenomenon that an adequate theoretical grasp of it has to be the result of a friendly division of labor and of a cooperative approach. Regarding, finally, the substantive merits of Feenberg's work, its empirical orientation that has rightly been praised by Achterhuis points us to actual and distinctive accomplishments.[9]

In the spirit of a common enterprise, I now want to take up a question that naturally arises from Feenberg's work: How ready are people in this country to engage democratic technology and how far are they willing to take it? My answer rests on a couple of interesting cases and thus constitutes anything but a conclusive answer. It merely raises issues and quandaries that point to the need for further research, some of it in the social sciences.

To begin, Feenberg himself has contributed important parts of the answer, and he has established at least this: ordinary (and extraordinary) people have resisted the utilitarian drift of technology more often than we realize. The efforts that are devoted to democratic technology often get in the way of maximizing aggregate consumption. If technology had been left to its own devices, total affluence in terms of GDP over population now and at least in the short term would be greater than it has been under the restraints of what democratic technology has achieved. But such technocratic success and prosperity would have come at the price of less freedom, greater inequality, and a worse environment.

How far are people willing or able to go in opposing technocracy? A crucial case in point is the governance of the economy, the place where the power of technocracy is concentrated. Feenberg reminds us that self-management was one of the explicit goals of the 1968 uprisings in France, and he thinks that it is in principle a feasible way of running a firm or factory. He says: "Of course the practical details were never settled, but it is at least plausible that the relatively well educated employees of an advanced capitalist country could substitute themselves for stockholders in guiding the policies implemented by the firm. Certainly, they would be better informed and closer to the issues than most stockholders" (40 and 42).

When in 1984 Rosauers Supermarkets, with a store in Missoula, Montana, were likely to go under, the employees bought the store "through a wage set-aside that averaged about 10 percent. The final payment was made in 1997."[10] But in the summer of 2000, the employees exchanged the privileges and burdens of management for cash. They sold the chain to a wholesale grocery distributor, and "the employee owners of Rosauers reportedly profited from their original $25.5 million purchase of the supermarket chain a decade ago."[11] Missoula's fabled Freddy's Feed and Read was originally owned by a cooperative, but more and more of the members wanted to be bought out. In the late nineties, the remaining owner struggled to keep the store afloat. Customers invested in the store to help. But finally, and in important part through the descent of Barnes and Noble on Missoula, Freddy's closed shop. Even the widely admired Mondragon Cooperative Corporation in Spain is struggling to preserve its principles of solidarity and cooperation in the face of the postmodern global economy.[12]

In 1994, United Airlines sold 25 percent of its parent company's stock to its pilots and 20 percent of the stock to the remaining employees to weather hard times and gain the employees' allegiance. The stock purchases were financed through wage concessions. Though the pilots were represented on the management side, contract negotiations in the summer of 2000 were difficult. The pilots did gain wage increases that made them the best paid pilots in the industry. In the wake of this agreement, United Airlines' other employees gained substantial increases as well, and as a consequence the airline had the industry's second highest labor costs and thus was among the most vulnerable to the shock waves of 9–11. Its stock plummeted, and, US Airways having gone bankrupt, United Airlines is the most likely airline to follow. Whether stock ownership by its employees will make UAL more resourceful than its competitors and resilient enough to avoid bankruptcy remains to be seen.[13]

What these stories suggest is that for workers the crucial problem of self-management is not expertise—Feenberg is right about this—but allegiance divided between the desire for self-determination and the desire for affluence. Hence, we may have to add to Feenberg's "ambivalence of technology" the ambivalence of people in technology-affluent democracies.[14] Workers in self-managed firms are divided in their roles as owners versus their roles as employees.[15] As owners they need low wages to have a competitive advantage and to collect profits for research and development and to keep up, renovate, or replace production facilities. As employees they want high wages. Individual firms today work in an economic environment that is not of their making and that they cannot control.[16] A self-managed firm can weather a difficult episode through self-exploitation. Workers as owners pay themselves as employees less than before or than desired or than what is paid elsewhere for comparable work. But you can work in the teeth of superior forces only for so long. If you do it for longer, bankruptcy and poverty ensue.

From Feenberg's standpoint, several reasonable answers are available. Democratic technology may not be a universal remedy for today's social ills. Perhaps we do have to let the global economy prevail in principle, but hedge it about with secondary instrumentalizations that blunt its social (and environmental) blades. However, we should not a priori exclude the possibility that democratic rationality can be brought to bear on the technology that is internal to the global economy.

One obstacle, then, in the path of democratic technology may be people's preference for affluence over autonomy. To the extent that the propensity of workers are divided between power and riches, the division is aggravated by the structure of capitalism. The fruits of the economy are typically divided according to the temporary equilibrium that the struggle between employers or management and employees or unions has reached. We find it hard to do in the macro economy what we must daily do in the micro economy of the household—strike a balance between savings and expenditures. Workers placing themselves resolutely on the expensive side of a corporation may not make more money, but they do simplify their positions and aspirations in the structure of the economy.

A second obstacle is the burden of engagement in technology that people are loath to shoulder even if there is financial gain. A widely hailed case of wresting common control from corporate commodification is the operating system Linux (sometimes called GNU/Linux) whose source code is open, freely available and modifiable, and cared for and improved through the voluntary efforts of countless programmers. Thus, everyone has the opportunity to gain the franchise and suffrage in a crucial region of cyberspace. Advocates and users of MP3 technology have similarly argued that music should be theirs to share as they see fit.

For Herbert Hrachovec, knowing Linux is more than taking advantage of an opportunity to democratize technology; it is very nearly, at least for philosophers, a moral imperative. He concedes that "GNU/Linux has a much steeper learning curve [than a Windows application] (though this is currently changing), but offers autonomy and sharing on a much broader scale. Self-determination, i.e. the ability of producers to own and adapt their means of production, can only be achieved in the second case."[17]

Charles C. Mann, a contributing editor of *Atlantic* magazine and a knowledgeable user of software, tried to take up the burdens of autonomy and live with Linux, but arrived at a decidedly ambivalent conclusion.

> Computers have become too important to be unquestioningly turned over to others to run. But they are a tremendous burden to maintain without help. Living with Linux has increased my awareness of the control exerted by computers even as it has stripped away my easy certainty about how much value I attach to liberty. Richard Stallman and Linus Torvalds may have launched a successful campaign to unseal every computer, but it is a pending question how many people will want to look inside.[18]

Linux is doing well and is about to become a major presence in the servers used by businesses.[19] But this is no longer a struggle of individuals and citizens with corporate commodification, but rather competition among major industrial players. Even the MP3 enthusiasts are beginning to pay for subscription services because the land of freedom has become treacherous. Techies approaching midlife and having more money than time would rather pay for a wide selection, high quality, and a sleek interface than engage in hand-to-hand combat with unreliable files, phony offerings, or hidden viruses.[20]

Pondering these several problems, we can see a fruitful research program emerging from Feenberg's theory of secondary instrumentalization and democratic technology. The first part is a comprehensive quantitative account of what standing democratic technology has achieved in contemporary culture. Next comes the description of the political, economic, and technical obstacles that stand in the way of advancing democratic technology. Third is the investigation of people's attitudes toward democratic technology and the cultural and moral hurdles it has to overcome. Finally, there need to be proposals for the advancement of democratic technology.

Although these four parts imply and benefit one another, each can be pursued by itself and at some length before it definitely requires convergence with the other three. In this vein, I will conclude with some suggestions on expanding democratic technology, suggestions that have been inspired by Feenberg's work in distance learning. He has pointed out that, after the austere pioneering stages he was importantly involved in, distance education inflamed the enthusiasm of the computer industry and of administrators in higher administration.[21] As a result, hundreds of millions of dollars have been spent on the development of colorful and artfully choreographed course materials that in part are available on CD ROMs. This development amounts to a kind of commodification, as David Noble has urged, and commodification and democratization are opposing forces, as Paul Thompson has argued in this anthology.[22]

The disaster that could have ensued for democracy and education was averted, not by providence and wisdom, but by the project's spectacular failure to interest and engage students. There will be special and limited needs that distance education can serve, but nothing more, unless the Armed Forces by virtue of huge expenditures force a semblance of success, not that the Department of Defense has invariably been a shrewd investor.

Feenberg believes that success in distance learning will not lie in the direction of further expense and sophistication—more color, more visual aids, more entertainment, more charismatic performers, etc. He thinks it may lie in the opposite direction, back to spare interactive text-based systems. It is a matter of taking up and developing a technology that faculty people have known for a long time and know how to use well, viz., texts and communication via texts.

There is a larger lesson for democratizing technology in Feenberg's proposition. The levels of engagement and appropriation in technology correspond to levels of manufacture and production. Total and radical appropriation would reach down to the most basic level of design and production. This was the dream of hippie homesteaders aspiring to self-sufficiency. They would heat their stoves and ovens with wood they would cut from their own wood lots, which they would stack and allow to season, and which they would split and carry to their oven on a chilly fall morning. They would wear sweaters made of wool sheared off their own sheep, carded, spun, and knitted by their own hands. And of course they would eat nothing but what they would grow and harvest or raise and slaughter themselves.

Such efforts were partial at best since no hippie community contemplated digging ore, smelting iron, and forging their tools. More damaging to the ideal of self-sufficiency, many a hippie had a bank account due to a former life or conventionally industrious parents. Computer technology makes it particularly clear that a radical posture is effectively self-defeating. Not even Hrachovec would suggest that we penetrate Linux down to the compiler language and ultimately down to the machine language. Thus, secondary instrumentation must be judicious in picking a level of technological construction that is deep enough to permit significant appropriation and change and high enough to avoid stultifying detail and allow for reasonable feasibility. Feenberg's approach implies that operating systems lie too deep and are too tedious.

The second lesson of Feenberg's distance learning proposal teaches us that democratic technology does not usually spring from a spontaneous upwelling of collective action. There is a need for leadership, but it has to be leadership that opens doors and engages effort rather than disburdens and disfranchises. Feenberg has exemplified such guidance and pointed out its fruits in his conception of distance learning. More important, his theory of democratic technology has opened a door and invited the cooperation of his colleagues in the philosophy of technology.

Notes

1. Parts of this chapter will appear as a review of Andrew Feenberg's *Questioning Technology* (London: Routledge, 1999), forthcoming in *Research in Philosophy and Technology*.
2. Larry Hickman, *Philosophical Tools for Technological Culture* (Bloomington: Indiana University Press, 2001), 157–84.
3. Langdon Winner, "Do Artifacts Have Politics?" in *The Whale and the Reactor* (Chicago: University of Chicago Press, 1986), 23–24.
4. Parenthetical page references are to Feenberg's *Questioning*.
5. Feenberg, "Do We Need a Critical Theory of Technology? Reply to Tyler Veak," *Science, Technology, and Human Values* 25 (2000): 238–42.
6. Lawrence Lessig, *The Future of Ideas* (New York: Random House, 2001).
7. Ibid., 26–48.
8. Jesse Tatum, *Energy Possibilities* (Albany: State University of New York Press, 1995); Gordon G. Brittan Jr., "Wind, Energy, Landscape," *Philosophy and Geography* 4 (2001): 169–84.
9. Hans Achterhuis, "Andrew Feenberg: Farewell to Dystopia," in *American Philosophy of Technology*, ed. Achterhuis, tr. Robert P. Crease (Bloomington: Indiana University Press, 2001), 65–93.
10. John Stucke, "Rosauers Gets New Ownership," *Missoulian*, 18 June 2000, D1.
11. Ibid.
12. George Cheney, *Values at Work: Employee Participation Meets Market Pressure at Mondragon* (Ithaca: Cornell University Press, 1999).
13. Edward Wong, "United Air's Family is Anything But," *New York Times*, 6 October 2002, section 3, pp. 1 and 11.
14. Feenberg, *Questioning*, 76 and 222.
15. John O'Neill, "Exploitation and Workers' Co-operatives: A Reply to Alan Carter," *Journal of Applied Philosophy* 8 (1991): 231–35.
16. They do influence and sometimes control the political environment through lobbying and campaign contributions. Thus, they help shape the economic environment indirectly and collectively.
17. Herbert Hrachovec, "Linux and Philosophy," available at www.ephilosopher .com/philtech/philtech11/htm on 26 February 2001.
18. Charles C. Mann, "Living With Linux," *Atlantic Monthly*, August 1999, 86. See also Alan Jacobs, "Life among the Cyber-Amish," *Books and Culture*, July/ August 2002, 14–15 and 40–42.

19. "Microsoft, Linux Gaining Momentum," Associated Press, available on 14 October 2002 at www.nytimes.com/aponline/technology/AP-AP-Tech-Conference.html
20. Neil Straus, "Online Fans Start to Pay the Piper," *New York Times*, 25 September 2002, section E, p. 1.
21. Feenberg, "Distance Learning: Promise or Threat?" available on 24 May 2001 at www.rohan.sdsu.edu/faculty/feenberg/TELE3.HTM.
22. David Noble, "Technology and the Commodification of Higher Education," *Monthly Review* 53 (March 2002), 26–40. Paul B. Thompson, "Commodification and Secondary Rationalization," this symposium.

CHAPTER EIGHT

PAUL B. THOMPSON

Commodification and Secondary Rationalization

This chapter presents a sketch of some ideas on commodities and commodification that I have been thinking about for a long time. I argue that an account of what I call technological commodification can move Andrew Feenberg's critical theory of philosophy along a step or two, and at the same time link it more firmly with the actor network theory of Latour, as well as with certain themes introduced by other philosophers of technology, including Langdon Winner, Albert Borgmann, Don Ihde, and Larry Hickman. In saying this, I will present my ideas on the direction in which Feenberg's theory should go, and it will be in this way that I provide Feenberg with some opportunities for critical response.

My theory of technological commodification owes an obvious debt to institutional economics. I will utilize ideas that have been developed in connection with theories of public goods, market failure, and allocative efficiency. Although they profess to be positive theorists, most economists have specific normative ideas about how markets are supposed to perform. I do not share many of these normative attitudes, and the use that I will make of familiar ideas from economic theory is, I believe, unique. My primary inspiration for this work comes from Douglas North[1] and A. Allan Schmid.[2] Both North and Schmid are more interested in law and policy (what I shall below call "structural commodification") than in technological change, though neither would find what I will say about material technology entirely foreign to their thought. In any case, the main themes of this chapter apply these economic ideas in a context (philosophy of technology) that *is* quite distinct from that of institutional economics.

Feenberg's Critical Theory of Technology

Andrew Feenberg's 1999 book, *Questioning Technology*, reorients philosophy of technology toward a proactive stance of engagement with technical professionals to pursue more democratic and rewarding design and implementation of technological systems. Much of the book is devoted to demonstration of the faults and errors in previous work on technology, including critiques of so-called essentialists, who see technology as alienation from nature, as well as of Feenberg's forerunners in critical theory, Marcuse and Habermas. Feenberg's positive theory building begins with a discussion of work by Bruno Latour and other "social constructionists," such as Michel Callon, John Law, and Wiebe Bijker. Feenberg develops three important themes from his discussion of this recent work on science, technology, and society (STS).

1. *Contingency*. The "meaning" of a given technology is not fixed in advanced by technical parameters (as both Heidegger and Habermas—with different goals and emphases—might have thought). The final, familiar form that technologies such as pasteurization, jet engines or the bicycle actually take depends as much on social networks and alliances as upon technical imperatives.
2. *Value-laden ness*. Tools, technical apparatus and technical capabilities (what we popularly call 'technology') are the superstructure that binds diverse human interests together in stable networks. But this binding is far from neutral. Values (such as particular conceptions of efficiency or property rights) become embedded in technical design. As such they both constrain and enable certain forms of individual and collective action. It is in this sense that Latour describes technical systems as 'actors'. It is also in this sense that technical elites are wrong to insist that technology is, in itself, neutral.
3. *Reform*. The hope and responsibility for normative, philosophical engagement with technology lays in attention to the way those technical systems are integrated into society and subsequently into the life experience of individuals. It is especially crucial to examine the interface between social networks and technical design. To the extent that this interface reflects open, democratic and broadly participatory interaction

> with all affected parties, technical systems will be progressive and nourishing, rather than oppressive and alienating.[3]

There are many details that need to be filled in and worked out with respect to the territory that Feenberg has already covered in making this argument. Yet I am more inclined to accept the basic validity of these three key moves in his program, and to press ahead with the theory building needed to implement his program of reform. Feenberg's own suggestions in this respect consist primarily in his theory of secondary instrumentalization. As distinct from primary instrumentalization, which seems to be the program of technical abstraction routinely conducted by applied scientists and engineers, secondary instrumentalization indicates the way in which tools, techniques, and technical systems become embedded in social institutions. One of Feenberg's key points is to describe a dialectical movement in the development and implementation of technology that recognizes a weakness that Habermas saw in Marcuse's thought, while nevertheless maintaining Marcuse's normative commitments to the reform of technology. Thus, primary instrumentalization corresponds fairly closely to what Habermas called "technical rationality," and proceeds largely as conceptualized by the technical disciplines. But this process does not result in a fully realized technology, requiring, in addition, *systematization*, or the establishment of connections to the existing natural environment, *mediation*, or the establishment of value-laden connections to the existing social environment, *vocation*, or inculcation into ongoing ways of life, and *initiative*, or integration into patterns of strategic interaction among competing interests. It is with respect to these latter elements that Marcuse's call for a new science, an alternative technical rationality, can be realized.[4]

To the extent that the theory of secondary instrumentalization is motivated by the Marcuse-Habermas debate, it, too, is backward looking. As developed in *Questioning Technology*, these four elements do not provide much guidance to people who hope to act upon the imperative for reform. Such action presupposes a conceptualization of the way that technical systems can be realized in more or less democratic, open, and participatory ways. Feenberg provides this primarily through the extended discussion of specific case studies in his earlier book, *Alternative Modernity*. Here, topics such as technology and the arts provide examples of technology's interface with concrete human projects. Extended discussions of the French Minitel system and the formation of an AIDs community around the fight for control of the research agenda provide models for the kind of engagement and network building that Feenberg envisions. Feenberg hints that the inherent contingency and local nature of such examples defeats the possibility of a more general and comprehensive theory of secondary instrumentalization, though this is not a claim that he makes explicitly. If so, then we must simply content ourselves with case studies. But in this chapter I want to explore the possibility of reorienting some fairly well-worn and traditional ideas in the philosophy of technology in the service of theory that will bridge the gap between the broad moves of *Questioning Technology* and the specific cases of *Alternative Modernity*. In the process, I hope to bring Feenberg's work closer to that of Albert Borgmann—a move that he may not welcome—and also to Latour's work and some enduring themes in Marxism—one that he surely will.

Commodities

Commodities are, in the most basic sense, goods that are routinely bought and sold. Commodification (or commoditization) is, thus, the transformation of something that is not bought and sold into something that is, or, to the extent that something can be bought and sold in varying degrees, an increase in the degree to which it is. The term "commodity" is also used in special and technical senses. In economics and finance, the term *commodity* is reserved for agricultural goods such as grain and pork bellies or manufactured raw materials such as steel or sawn timber. What characterizes such goods is that they are graded and standardized in a manner that allows any one specimen of the good to be regarded as a perfect substitute for another. As such, commodity markets are quite unlike consumer markets for houses or automobiles. When most of us purchase a home or automobile, we purchase a definite, particular house or car, but a commodity purchase can be satisfied by delivery of any lot of the good in question meeting the requisite quality standards.

The suggestion that commodities and commodification are, in some sense, bad or regrettable is an enduring theme in leftist social theory. Some infer a capitalist mode of production from the use of the term *commodity*, and the term is often associated with stigma or some sense of negative valence. The main theme, of course, is the claim that the institution of wage labor makes human productive activity into a commodity, resulting first in the forms of alienation articulated in Marx's *1844 Manuscripts* and also in the contradictions of capitalist political economy that he exposited throughout his mature work. Here I will not be concerned with the interpretation of Marx's writings, for my goal is to articulate a conceptualization of technology's role in commodification that is analytically independent from them, however much it remains indebted to Marx's inspiration and larger philosophical projects. It is clear that technology is in some manner *related* to commodification in traditional Marxism, if only in that the rationalization of the labor process described by Adam Smith is thought to create the conditions for a system of production that calls for wage labor. Subsequent theorists have linked technology and the commodity form even more closely. Albert Borgmann, for example, distinguishes commodities from things. For Borgmann, things demand our involvement and attentiveness, and as such structure human ends. Commodities, on the other hand, are unencumbering and available for use without thought or engagement.[5] In such contexts, commodification clearly connotes degradation or reduction in the value or dignity of goods. Such reduction in value or dignity often stems from commodified goods lacking unique or distinguishing traits that would have been characteristic of things or handcraft goods they replace.

As already noted, the term *commodity* is used in a more limited sense in economics and finance, where commodities are a special class of material goods that are traded under the peculiar conditions of commodity markets. The growth of such markets is associated with a special, though important, set of problems in social theory (and certainly for social theories indebted to historical materialism), namely, the decline of agrarian or feudal social relations and the eventual rise of industrial agriculture. Although I would hope that what I will say about commodification will enlighten and inform some of the contexts in which the term *commodity* is given a special or technical sense, I will not presume any specific technical sense for the term *commodity* in the following analysis. For present purposes, the simple definition of "something routinely bought and sold" will suffice. However, since the degree to which various specimens of a good can be interchangeably substituted affects the degree to which they can be bought and sold, I will end up with a view that recognizes the economist's special use of the term *commodity*. Oats and automobiles are both commodities, in my view, but oats, barley, pork bellies, and steel are goods that satisfy the commodity form more completely than do the houses and automobiles that ordinary people buy and sell every day.

Two Types of Commodification

Commodification occurs when goods and services are transformed in one of four possible ways. (1) Alienability (the ability to separate one good from another, or from the person of a human being) is altered. (2) There is a change in excludability (the cost of preventing others from use of the good or service). (3) There is a change in rivalry, or the extent to which alternate uses of goods are incompatible. (4) Goods are standardized so that there is an increase in the degree to which one sample of a given commodity is treated as equivalent to any other sample. Goods that exhibit the commodity form tend to be alienable, excludable, rival, and standardized. These four parameters constitute the ideal type for the commodity form, and much more will be said about each below.

There are at least two distinct processes of commodification that can be identified. Sometimes a good that has not been thought of as something that could be bought and sold comes to be thought of as a legitimate object of monetary exchange. Marx's famous discussions of wage labor provide the paradigm example of such a transformation, though there are many others. I will call this type of transformation *structural commodification*. In such cases, there has not been any direct change in the good. Instead, there have been changes in legal rules, customs, or morals that eventuate in practices of exchange that had hitherto been regarded as suspect or had not existed at all. Structural commodification occurs through the legitimation of alienation, exclusion, rivalry, and standardization as forms of social practice, or through the protection of such practices through coercion and the threat of force. The mechanisms of structural commodification involve political power, and the actors are, thus, agents capable of wielding or exerting such power.

The other process of commodification does involve transformation of the good in question, or of the material circumstances for exchange of the good. The invention of sound recording, for example, transformed musical performances into goods that are bought and sold many times more frequently than in Mozart's day. I will call this second type of transformation *technological commodification*. In technological commodification, alienation, exclusion, rivalry, and standardization are built into the goods, or one could say that the goods themselves are being materially transformed. Sound recording permitted the alienation of musical performance from the person of the artist. More detailed examples of each form of technological commodification will be given below. The mechanisms of technological transformation must involve some elements of physical causation, yet they certainly involve aspects of human agency, as well. Much of the recent work in science studies can be understood as illuminating these complex mechanisms. Bruno Latour, in particular, has argued for a principle of symmetry, that technological artifacts themselves are actors exerting influence on human beings.[6] The theory of technological commodification can be understood as a hermeneutic of symmetry in Latour's sense.

There is a tendency to think of commodification generally in terms of structural transformations that reflect shifts in the balance of power among various social interests. While labor had been hired out on a piece rate basis for centuries, the institution of wage labor reflects a commodification of labor power because the routinization of hourly wage rates suited the interests of manufacturers. It not only simplified and rationalized the factory owners' business practices; it drove down the total cost of manufacturing labor by placing workers into competition for wages. In such cases, commodification is associated with exploitation and reflects an imbalance of political and economic power. Here, it is easy to link commodification with a moral and political critique.

This is clearly the form of commodification on which Marx himself was most eloquent, yet he was also quite aware that the rationalization of the labor process was also a process of technological innovation. Although any given factory manager would choose to implement such innovations in accordance with their economic and political interests, the possibility of such innovations places capitalists into competition with one another. This competition among capitalists unleashes a technological treadmill: everyone runs harder just to keep up. At the same time, technological changes are generally thought to reduce prices for consumers. This has, for many liberals and neo-liberals, made processes of commodification that occur through technological innovation seem less problematic from a normative standpoint. One motivation for theorizing technological commodification is to clarify and articulate the normative significance of alternative technology.

Although structural and technological commodification differ in obvious ways, there will be cases where the distinction between them may be a matter of interpretation. This will especially be the case when technological transformations occur in the method of exchange. For example, grains were once routinely transported and sold in bags. With the advent of railroads and grain elevators, there was a total revolution in the way that grain was bought and sold, and it is quite appropriate to describe the accompanying shifts in power and market structure as instances of commodification. On the one hand, it was still lots of wheat, oats, and barley that were being bought and sold, and only the market structure for exchanging them was different. The change in market structure demanded changes in attitudes and the development of grades and standards for judging that one lot of wheat, oats, or barley was equivalent to another. As such, it may be appropriate to interpret the nineteenth-century rise of commodity markets for grain as an instance of structural commodification. On the other hand, this was a manner of exchange that was entirely unfeasible before the advent of rail and elevator technology. As such, it seems equally appropriate to interpret the nineteenth-century rise of commodity markets for grain as an instance of technological commodification. A further argument for the latter claim is made below.

In any case, the purpose of calling attention to the technological dimension of commodification is not to deny or minimize the role of economic and political power. Following Latour, the mechanisms of science in action should be understood as a complex mixture of different forms of agency, some of which are of a familiar political sort.[7] This creates what Don Ihde calls a multi-stable phenomenon, meaning that a technical artifact is open to more than one interpretation, but also that some avenues of interpretation are more stable, more capable of supporting social linkages, than others. Reductively emphasizing sociopolitical mechanisms opens into an interpretive domain that has been emphasized by those who wish to call attention to the role of political power. Yet in Marx's own analysis, capital has the kind of power that it has because of peculiar features of industrial technology. Reductive emphasis on these features leads to technological determinism.[8] Latour and Ihde are calling for us to resist both forms of reductionism, and to keep complexity in play. The theory of technological commodification provides a hermeneutic framework in service to such complexity, and as such yields a firmer basis for developing a critical theory of technology. To that end, it will prove useful to clarify the role of technology in effecting each form of commodification.

Alienation

Philosophers routinely distinguish between a psychological or existential sense of alienation and a political-economic sense. With respect to the latter, the term is often used to indicate a distinction between alienable and inalienable rights. Alienable rights are those which can be transferred from one rights holder to another by sale or gift, while inalienable rights are those which are thought to be inherently nontransferable. Classically, alienable rights are property rights, including the rights to occupy, use, and profit from the use of a given item, while inalienable rights include basic liberties, including an individual's right to life. The basis for this distinction is not always clear, however. For example, Jefferson's famous reference to the inalienable rights of life, liberty, and the pursuit of happiness in the American Declaration of Independence is generally thought to stipulate a moral claim about the injustice of any usurpation of these rights. Garry Wills, however, argues that Jefferson is simply making a descriptive claim. These are rights that are inherently tied to the person. They are nontransferable, and since everyone has their *own* rights to life, liberty, and the pursuit of happiness, there would be little benefit in transferring them were it possible to do so.[9]

Wills apparently thinks that one's rights to life, liberty, and the pursuit of happiness are no more separable from one's lived body than are one's subjectivity and point of view. Such rights can be *violated* but not *alienated*. This way of construing alienation suggests a closer link between the psychological conception of alienation and the economic one than is generally thought. For under a system of wage labor, a worker may alienate her right to profit from her labor, and may sell it to the capitalist for an hourly wage. But if the Marx of the *1844 Manuscripts* is correct, this form of alienation—which would not have bothered Jefferson, by the way—is also a psychological and metaphysical alienation. To say that the institution of wage labor commodifies the worker through a process of alienation is thus descriptively correct, but to interpret this as a moral claim may require additional argument.

The issues are not unique to the alienation that occurs in establishing an institution of wage labor, however. As already noted, recording technology permits the alienation of a musical performance from the lived body of the performer. This technology thoroughly transforms the social and material character of musical performance, though the description of this transformation can be contested. Does recording technology create a wholly new commodity good, or does it commodify a good that previously would have required a more complex relationship between performer and listener? Borgmann, for one, prefers the latter account, and argues that while recorded performances are wonderful things, they involve an attenuation of music as a form of social bonding, community, and shared meanings.[10] For my purposes, I would prefer to treat the normative evaluation of commodification as a distinct issue (to be taken up briefly in the concluding section). What is critical about Borgmann's observation is that the social relationships between hearer and listener of recorded music can revolve primarily around a series of monetary exchange relations needed to produce the recorded performance as an alienated good. Sound recordings also involve each of the other dimensions of the commodity form, and it is difficult to distinguish their respective contributions to the commodification of music. Yet without the technologically produced alienability of the good it is, I hope, obvious that any growth in commodity relations would have to take a very different path.

Further evidence of commodification can be provided by noting the nature of normative questions that arise once the musical performance has been alienated from the person of the performer. What are we to say about the right to use or profit from performance? Is it clear that the performer should retain the right to limit or control the use of the performance? Is it permissible for someone who records a performance to transfer that recording to a third party, or does the original performer still have a say? Are there any limits on the benefits that a person in possession of the recording can derive from its use? In particular, does someone who possesses a recording have the right to profit from it? The questions themselves testify to commodification in their unrelenting focus on profit, compensation, and monetary concerns.

Technological alienation need not involve alienation of a good from the human body. Recent discoveries in biology allow scientists to alienate plant traits from the seed stock with which they have been inalienably associated for millennia. Again, many of the key normative questions following on the growth of biotechnology testify to the commodity character of the new goods—plant traits keyed to a genetic sequence—that have been alienated from an old commodity good, namely, seeds. Does the ability to isolate a strand of DNA give one the right to control the use of that DNA in other plant species? If one alienates a medicinal property from one plant and transfers it to a different species, has one established a right to control the use of that trait, or to share in the profits (or other benefits) that others derive from using it? These are the classic questions that are raised by any act of commodification. Because a society lacks norms or ways of thinking about the use and exchange of things that it was hitherto impossible to treat as separable commodities, the vacuum of established or legitimate practice is often filled by practices that reflect immediate short-term interests of empowered parties.

Technology can also affect the alienability of a good from a particular place or situation. Labor historian E. P. Thompson's seminal article entitled "The Moral Economy of the English Crown" describes a key transition period in the transition to capitalism. Prior to the thirteenth century, village economies in rural England were based on a system of commodity exchange. Village farmers would grow and harvest grain. Millers were entitled to a share of grain milled for their services, as were bakers or other village merchants, and, for that matter, landlords. This system went into collapse after improved roads and wagons made it possible for farmers to transport grain from village to village, seeking the most attractive terms of trade. Outraged villagers rioted, arguing that farmers had no right to do this. Local grain growing in the fields belonged to the village in common, they claimed, with each villager playing a contributing role in the common production of bread. As Thompson shows, the English crown did not support this view, ushering in a new era of commodity exchange and property rights. The legal validation of farmer (or landowner) property rights was an example of structural commodification, but the dispute had never come up before better transport technology permitted alienation of harvested grain from the local village setting. Prior to that technological innovation, grain was not even a candidate for being a commodity good.[11]

Thompson's example suggests that transport and communication technologies probably have sweeping influence on the way that a variety of goods and services do or do not participate in the commodity form. The ability to easily and cheaply alienate a good or service from the place or situation in which primary production occurs will certainly facilitate the opportunity to seek attractive terms of trade. Goods or activities that are not so easily alienated may not even be conceptualized as possible items for monetized exchange. It is a commonplace to notice that modern transport and communication technologies contribute to the expansion of markets, but it is worth noting as well that they can also affect the degree to which our interactions with one another (and also with nature) are dominated by the possibilities and moral questions of market exchange. In saying that these technologies affect our interactions with one another, we are describing technologies as actors in Latour's sense. By bringing about a systematic alienation of goods from specific places, these technologies resituate not only the goods themselves, but also the human actors who participate in exchange. Since it is doubtful that any human agent could have foreseen the full range of consequences associated with such transformations, it is misleading (in the sense of being reductive) to suggest that the intentional acts of human beings are solely responsible for them.

Exclusion

Invention of alienability is probably a fairly rare occurrence, though one with far-ranging consequences when it occurs. A more common form of commodification occurs when technology reduces the cost of excluding others from access to the good in question. Classically, goods such as air, sunshine, and water are not thought of as commodities simply because they are widely available, and because in many instances it would be difficult and costly to keep people from getting them. It is possible only in places such as deserts, where water becomes a commodity, not only because it is scarce, but also because it is relatively easy to control access to springs and oases. Power to exclude may reside in the legal structure, and codifications resulting in new powers of exclusion represent classic instances of structural commodification. The enclosure movement in Britain and across Europe transformed agricultural production from a feudal mode of production to one of industrial capitalism, and the legal innovations in property rights that facilitated it are an instance of structural commodification.

However, commodification by exclusion can occur by technological means. One method is through a direct transformation of the commodity good itself. The invention of artificial lighting makes light into a good that can be produced at will in a particular place, and as such, a good to which access can be somewhat controlled. More typically, however, a good is commodified by exclusion through technology that affects the conditions or environment in which goods are used or transferred. Fences are the classic technology of exclusion, and technological innovations that reduce the cost of fencing serve commodification. Although the enclosure movement was a complex transition involving sociostructural transformations, a key material transformation consisted in the cost of excluding commoners from the most fertile lands. The development of barbed wire facilitated the creation and enforcement of the commodity form for rangelands across the Americas. Here the legal structure of property rights was already in place, but customary rights of access permitted transit across private lands. Landowners bent on exercising legal rights against trespassing faced difficult odds. Lands cheaply fenced in barbed wire had a dual effect: they created a physical barrier that made the act of trespass itself more inconvenient, and they inscribed legal property rights onto the landscape in a way that made enforcement of the law (and proof in court) more cost-effective. Right of access itself became a commodity that could be negotiated. In a manner of speaking, then, all the technologies of security and coercion serve commodification in this sense.

But technology can also feed a process of decommodification by increasing exclusion costs. The Xerox machine and the many technologies of copying information in magnetic or digital form provide excellent examples of decommodification. The ability to protect property rights in recorded performances, written words or visual representations presupposes that these goods exist in the commodity form. As the excludability of information-based goods declines, so does their participation in the commodity form. As such, those who own property rights in information-based goods must move to recover the lost excludability in the form of patents, copyrights, or other legal norms that depend upon the police power of the state, rather than on the characteristics of the goods themselves. This places the legitimacy of their power to exclude into question, as the current debate over Napster and the MP3 format for recorded music attests.

Changes in exclusion costs that weaken the commodity form provide excellent examples of unintended and even unimagined impact from technological innovation. Certainly the inventors and early developers of recording and information technology did not intend the decommodification of recorded performances or written texts. Again, they exemplify the way that Latour understands technologies and techniques as actors: "Why not replace the impossible opposition between humans and techniques by association (AND) and substitution (OR)? Endow each with a program of action and consider everything that interrupts the program as so many anti-programs. Draw up a map of alliances and changes in alliances."[12] Exclusion is generally part of the program of fences and will put them into an alliance with humans who will benefit from commodification. Inventors of information technologies may have been profit-seeking capitalists, but by reducing exclusion costs, their progeny have bolted into an alliance with other human actors who have no interest in preserving or expanding the commodity form.

Rivalry

Commodity goods not only tend to be excludable, they tend to be rival. That is, multiple uses of the selfsame good cannot be made. This is a slippery notion that can mean a number of closely related things. First, rival goods are consumed or "used up"; they do not endure or persist for multiple uses. If I eat a peach, it is gone. Books are less rival than peaches in this sense, because a book can be read over and over again. Second, rival goods can only be accessed by one user at a time. Web pages or billboards are less rival than books in this sense, because any number of people can access the information on a billboard or Website at the same time. Finally, rival goods do not support multiple types of use. A bean is less rival than the automatic transmission from a 1956 Buick Roadmaster because I can use it for a burrito, a seed, a projectile, as stuffing for a bean bag, or as a token in any number of games. The Roadmaster transmission can serve as a rather specific piece of automotive equipment or as a generic heavy object, but that's about it.

Greater rivalry increases the degree of commodity form because people who need rival goods must participate in exchanges to acquire them more frequently. In market economies especially, the degree of rivalry among goods affects the frequency with which people find themselves participating in market transactions, and as such the degree to which they are bound to the market as an institution. This affect is clearly exhibited in a recent example of commodification through increasing rivalry: the development of the Terminator Seed. So-called Terminator Seed has been genetically transformed so as to produce infertile progeny. Farmers can grow normal, high-yielding crops from Terminator Seeds, but the seeds that they save from that crop will not germinate and produce a crop again in the succeeding year. This means that farmers must return to seed companies to purchase seeds year after year. A Terminator Seed is more rival than a normal seed because growing it once "uses it up," and the farmer must return to the seed company in the same way that I must go back to the grocer when all of my peaches have been eaten.

Reducing rivalry is also a form of decommodification. Again, information and Internet technologies provide good examples. As already noted, a Website is less rival than a book. Note, however, that a Website can be made very excludable by protecting it with a password. The information on the Website would nonetheless be relatively nonrival, since it can be accessed by many people again and again. Of course it might be possible to *increase* the rivalry of the Website by issuing passwords that offer only one-time use. Thus, the net effect of technologies on the commodity form may not be particularly straightforward. From the perspective of an author or a publisher, information printed in books is rather nonrival. The books can be given away or sold in used book stores, and that reduces the commodity character of a written text. If books are displaced by open access Websites, the commodity character is dissipated further. But if books are displaced by password protected Websites, the increase in exclusion cost makes the written text more of a commodity from the perspective of an author or publisher, while less of a commodity from the perspective of a shopkeeper. If the passwords expire after one or two uses, the result (I think) is a net increase in the commodification of the written text, since it is at least as excludable as a book, and far more rival.

Again, Latour's language of actor networks lends itself well to the transformations and shifting alliances that such technological innovations involve. The shopkeeper is enrolled in an alliance with the technologies of printing and bookbinding, and in the traditional state of affairs, authors and publishers are also part of the program. The printed book represents a blend of alienation (the text is alienated from the voice and the hand), exclusion (the text is only available to those who have the book), and rivalry (the text may be used over and over by anyone who has the book). This blend enrolls authors, publishers, and shopkeepers into a network along with the books themselves (as well as the tools of printing and bookbinding). Then the password protected Internet technology appears on the scene with a new blend of exclusion and rivalry (the alienability seems to be unaffected). This amounts to an antiprogram such that author and publisher may shift alliances and a new network is formed. Whether this change of circumstance is a good thing or a bad thing is a further question, but it is not neutral. It is not as if (to use Ihde's phrase) the book and Internet are just pieces of junk lying around until authors, publishers, and shopkeepers decide to use them in a particular way. Capabilities for exclusion and rivalry are materially instantiated in these technologies, and these capabilities make them active within human affairs.

Standardization

Classic commodities such as corn, oats, and cans of Campbell's condensed tomato soup have the peculiar feature that makes the Chicago Board of Trade possible. One lot or sample of the good can be characterized according to quality and quantity standards that make it perfectly substitutable for any other. It is worth revisiting the example of nineteenth-century grain-handling technology once again because many arguments about industrialization turn on the issue of standardization. Clearly, the manufacturing technology of the middle industrial period depended heavily on standardization. Eli Whitney, for example, is cited as a founding figure of industrialization because he was able to introduce interchangeable parts into the manufacture of munitions. This replaced a craft industry with assembly line technology, "rationalized" the labor process by "deskilling" it, and was instrumental in creating the mass markets for low-paying wage labor jobs. For more than a hundred years, it seemed as if technology contributed inexorably to commodification through a process of deskilling that was implicit in the process of standardization.[13]

For perhaps two hundred years, technologically induced standardization of goods served commodification both directly, by turning handcrafted goods into standardized commodities, and indirectly, by increasing markets for wage labor and squelching tastes for made to order goods. But automation technology may now have reached a point where it will serve the made to order good, and this may be reversing the flow somewhat, introducing yet a third type of decommodification into capital markets. While this may reduce the drab sameness that we often associated with mass manufactured goods, the relationship between deskilling and commodification is complex. On the one hand, mass manufacturing is associated with standard or uniform goods. On the other hand, assembly line techniques and standardized parts allow work that might have required skill and craft to be done by untrained workers. This kind of deskilling is clearly related to the commodification of wage labor, and it continues to be an important feature of industrial relations today. In the last few decades, meat packers have broken longstanding union agreements by introducing powered cutting tools that allegedly permit unskilled labors to disassemble a carcass.[14] Here there is a technologically based standardization of the work process that may or may not be associated with interchangeable or standardized goods. Modern assembly lines use computers to "destandardize" the products, so that each automobile rolling off the line may be virtually unique. This technology may eventually go some distance toward relieving the monotony of choice among consumer goods, but it is not having any necessary affect on the standardization of the work process, or the commodification associated with wage labor.

Yet the creation of uniformity in goods is also a form of commodification. Standardization can make goods fungible in a manner that serves the growth of commodity relations. This is, for example, the case for wholesale marketing of standardized grocery products such as pickles or cans of condensed soup. One truckload is as good as any other, and the markets that develop for these goods reflect the intersubstitutivity of these goods. The extreme case, however, is for pure commodities. For traders, classic commodity goods such as oats, barley, and frozen orange juice concentrate are liquid assets whose entire purpose consists in their ability to be bought and sold. Their price is a function of what others are willing to pay, which is a function of the perceived scarcity of these goods far more than it is their final use value. At the end of the day, someone will buy these goods in order to fix breakfast, but this hardly enters into the meaning that oats, barley, and orange juice have in modern commodity markets. There, future contracts to deliver these goods will be traded for every manner of financial instrument over and over again, long before anyone with any material interest in using the goods themselves enters the picture. Such trades depend on one quantity lot of the commodity in question being completely fungible—for all purposes of exchange identical and intersubstitutable—with any other lot of equal quantity.

As William J. Cronon makes clear in *Nature's Metropolis*, the standardization that makes these commodities into fungible lots is a technological achievement. The oats, barley, and orange juice themselves may have differences in quality that would, under different circumstances, create differences in price. Standardization is possible because milling companies have the technological capacity to measure features such as kernel size and moisture or acid content, then to mix different lots. High quality grains are mixed with lower quality so that a uniform quality or standard grade is the result. The creation of this capacity for standardizing grains was a byproduct of nineteenth-century technology developed to offload, store, and reload grain during shipment from the North American Midwest to population centers in the East. Once the capacity for standardization existed, the creation of commodity markets in Chicago followed. The financial and political power of the grain companies abetted the structural transformation that created the commodity future as an exchangeable property right, but that transformation would not have been possible without the technological ability to standardize goods that are inherently nonstandard in their natural form.[15] Structural commodification—the creation of the Chicago Board of Trade—was accompanied by and would not have been possible without the technology of standardization. It is not at all unreasonable to use Latour's terminology in describing transport, loading, and storage technologies as actors in this transformation.

Commodification and Secondary Instrumentalization

I have introduced a new vocabulary for theorizing the process of commodification and thereby for characterizing the commodity form. It is intended to articulate a number of points that previous authors have associated with commodities and commodification, though of course these terms are used so broadly that this cannot be done in every case. The advantages of the new vocabulary are an increased capacity to map the complexities of commodification and decommodification, and in a clearer way to express how technology and technological innovation affects those processes. Although Feenberg himself does not base important claims on commodity arguments, this new vocabulary does facilitate a number of Feenberg's philosophical aspirations while linking his program both to social studies of science and to more traditional work in Marxist social thought. The theory of technological commodification will also permit us to argue that technological innovations call for the same type of justification and legitimization that we would normally demand with respect to constitutional, legal, or customary changes in social practice that affect the fundamental norms for claiming property rights or for regulating exchange. The nature of the normative arguments that I (and Feenberg, I hope) would like to make is discussed briefly in the concluding section of the chapter. The theory of technological commodification can be integrated into Feenberg's philosophy of technology as a more detailed account of secondary instrumentalization in some of its most influential and enduring forms.

As noted above, Feenberg illustrates this concept primarily through two case studies. One deals with AIDS research. Feenberg's treatment of this case is primarily prescriptive advocacy of a return to the notion that medicine should involve care of the sick, as opposed to today's emphasis on the technical ideal of curing disease. As characterized in *Alternative Modernity*, the issue here is an ideal of research that stresses experimental and statistical notions of control. Scientists working under this model regard those who wish to participate in experimental studies as useful up to the point that these technical goals are satisfied, but also as vulnerable to exploitation and false hopes for cure. When medicine is seen in terms of care without hope of cure, the nature of participation in research trials itself changes, as participants express solidarity with others (including those not yet afflicted with AIDS) and derive benefit from caring interaction, rather than the hope of cure.[16] As recounted by Feenberg, the AIDS case illustrates a point dear to the heart of first generation critical theorists: the scientific attitude and scientific research methods become determining factors in framing human relationships. Yet the case can also be interpreted through the hermeneutic of commodities and commodification.

Whether focused on caring for the sick or curing disease, medical care is, in some ways, a paradigm example of the noncommodity good. In its idealized sense, medical care is delivered through face-to-face interactions, rich in meaning and responding to particular circumstances of need and opportunity, rather than conditions of commodity exchange. This is not to say that delivery of medical care can ever be stripped entirely of its economic dimension, but in lore if not in fact, doctors and caregivers deliver care without regard to the patient's ability to pay, and receive compensation in many nonstandard and even nonmonetized forms. Think, for example, of the country doctor of a bygone era who accepts a ham or a supply of eggs instead of a monetary payment. However, it is also very clear that against this ideal, medical care has undergone a continuing process of commodification for at least one hundred years. In large part, this has been a structural commodification, as the rise of institutions such as health insurance, HMOs, medical specialties, and government health benefits have combined with capitalism's general tendency to frame social relations in the commodity form.[17] This social movement is generally characterized in terms of structural transformation. Technology has played a supporting role, however. The pharmaceutical industry is dedicated to the commodification of health care delivery, for example, alienating key therapies from the person of the physician, standardizing them and offering them in the form of highly rival pills having relatively low exclusion costs (especially when formulae are protected by patents). Furthermore, the rise of technically based medical specialties has certainly differentiated medical practitioners from one another in a manner that results in each delivering standardized commodity services that can be categorized and compensated on a market basis according to the requirements of bean-counting technocrats in hospital, insurance company, and government agency accounting departments. Feenberg's call for a return to the care of the sick is actually a call for the decommodification of medicine.

If the above analysis is correct, Feenberg's call might be answered either by conventional social reform of policy and practice (structural decommodification), or it could be abetted through technical design. Drawing on empirical work on social networks that emerged as an offshoot of political struggles over AIDS research, Feenberg's discussion of the AIDS community illustrates how researchers, drug companies, and clinical practitioners were conducting key elements in primary instrumentalization—the scientific research and development needed to draft a technical response—with the expectation that it would be shaped by these commodity-oriented dimensions of health care under late capitalism. That is, care would be delivered through drugs and therapies administered by specialists. Patients would, in the fullest sense of the commodity form, be consumers of products designed and implemented according to the primary technical imperative of control over disease processes and the secondary technical imperative of delivery according to commodity exchange. Instead, networks of AIDS sufferers, fundraisers, and scientists formed in a way that restructured the secondary implementation of drugs and therapies. Networks formed to ensure that patient needs were reflected in research priorities actually evolved into communities of care and support for AIDS sufferers, eventually delivering something much closer to the idealized conception of medical care as a noncommodity good. New drugs are a primary tool for AIDS therapies, but they continue to be administered through much richer human institutions than simple commodity exchange. Significantly, technical expertise has evolved in relatively nonrival and less standardized fashion, requiring (and in turn fostering) the support communities that provide such a key element of the AIDS story in Feenberg's treatment.

Feenberg's second case, the French Minitel system, is a more straightforward example of technological commodification. Feenberg's work on the French Minitel was first published in 1992. Since then the emergence of the Internet creates the need to revisit the details of the case, but much of what he discusses relates to the commodifying and decommodifying effects of information technology. Voice communication technology accomplishes a stunning alienation of communication practices from settings of spatial or co-presence, setting the stage for a new good capable of being delivered through market exchange. Nevertheless, telephone systems resisted certain aspects of the commodity form in their early years owing to the need for a large technical infrastructure that did not easily lend itself to commodity exchange. However, by the mid-twentieth century telephone services were being delivered as rival goods with relatively low exclusion costs. Developers of the Minitel, again, expected to use the expanded medium to expand their offerings for market exchange. However, the ensemble of communication technologies that have emerged with computer technology have made it vastly more expensive to maintain exclusive control of these services, even as they have created opportunities for multiple uses and reuses of network services. They have also spawned new goods through multiple transformations of communication practice, opening the possibility of new forms of commodity exchange. Although any adequate telling of this story would be a book in itself, it is nevertheless plausible to hypothesize that the revolution in information and communication technology—especially as told by Feenberg in the Minitel case study—can be characterized as a massive and primarily technologically generated decommodification of the systems that had been put in place during the first three-quarters of the twentieth century.

Loose Ends Connected, Norms Stated

In the interests of clarity, the examples from *Alternative Modernity* have been discussed without introducing the Latourian terminology of technological actors and their shifting alliances with human agents. Yet, hopefully, readers can supply for themselves how the story of AIDS drugs and the Minitel could be readily retold in exactly those terms. Feenberg himself has stated that the social studies of science provide an important reorientation for philosophy of technology. Yet his insistence on the importance of social relations and his utter disdain for all forms of technological determinism create some tension with Latour's principle of symmetry, which demands that techniques and tools be described as actors capable of enrolling their human counterparts into programs of action. There is a chasm, in fact, between primary and secondary rationalization. The former is simply pursuit of technical efficiency through conventional engineering means, while the latter often seems to consist in disentangling scientists and engineers from social networks organized around capitalist programs. It is possible to read Feenberg as one might read Marcuse:

denying any sense of efficacy to technology itself and seeing the entirety of mechanisms affecting the deployment of technology in terms of capital and the pursuit of profit.

Yet many passages in Feenberg's books disavow any intention to engage in what is, in effect, an essentialism of socioeconomic forces. As presently developed, his philosophy of technology is not equipped with ideas that allow us to understand technology itself as active in shaping human intentions and human relationships, and this omission limits the rapprochement with science studies. The theory of technological commodification provides a way to connect a few more dots between Feenberg and Latour. Technologies are active in commodification because the effects of alienation, exclusion, rivalry, and standardization exceed the intentions and expectations of the human actors—which includes researchers *and* capitalists—enrolled in the program. The theory is also suggestive of ways in which human actors can be attentive to technology at the basic stages of research and development. Technologies will not only do things that are already being done more efficiently; they will likely have wide-ranging unintended effects on the way human beings interact with one another. *Some* of these effects are mapped by commodification. The theory of technological commodification thus provides a set of questions that human actors can pose to the technologies with which they are proposing to become involved. Such questions can, at least in principle, alter the secondary instrumentalization of a technology much earlier in the research and development process. Thus, at both the level of describing processes of rationalization and at the level of reform, attention to the mechanisms of commodification extend and enrich the critical theory of technology.

The theory of commodification also paves the way for two kinds of normative argument. The first echoes longstanding leftist political philosophy, and has been argued most forcefully with the philosophy of technology by Langdon Winner. Winner has long argued that technologies themselves "have politics," and that material features of tools and techniques form "the technological constitution of society." Winner's own work has stressed the contrast between technological systems that require large physical and managerial infrastructure, such as centralized electric power generation, with those that support less bureaucratized and command-oriented control systems, such as alternative energy technologies including wind and solar power.[18] Feenberg takes pains to dissociate himself from the technological determinism implied by Winner's position, but he clearly supports Winner's view that in adopting a given technological system, a society makes normative political commitments regarding the institutionalization of authority, processes of decision making, and cultural diversity.[19] How are we to read the relationship between Feenberg and Winner on the technological constitution of society?

One possibility is that the secondary instrumentalization of technological systems should be understood exclusively in terms of what I have called structural transformations: changes in laws, policies, and customary norms. Such a view would permit Feenberg to agree with Winner's call for a democratization of technology, yet to deny that artifacts have politics beyond the way they are contingently and historically associated with particular groups of social actors who have politics. The same technology with different social allies has a different politics. However, the theory of technological commodification allows us to characterize alienability, exclusion, rivalry, and standardization as material features of the technologies themselves, wholly apart from any intentions or aspirations that their developers or other social actors have for them. But legislation and policy themselves address alienability, exclusion, rivalry, and standardization in social relations through mechanisms such as property rights, market structure, and regulation. No one questions the political nature of an attempt to alter given conditions of alienability, exclusion, rivalry, or standards through legislation or policy. Why should technological changes that bring about exactly the same sort of alteration in social relations be exempt from political debate?

I take it that the main normative claim made by those who want to democratize technology is simply to get these questions on the table. The theory of technological commodification should, if Feenberg accepts it, move his position closer to Winner's in this respect. Once technologies are recognized as political actors, it will be possible to take a number of different positions on whether changes in social relations effected through technological commodification are a good thing or a bad thing. A traditional conservative may be concerned about the way that such changes destabilize the existing social order. Thus, a conservative response to a technologically based decline in exclusion costs or rivalry may be to propose new laws that maintain the existing power relations. Proposals for laws to preserve the system of market relations favoring the recording industry and media holding companies are conservative in just this sense. As a Leftist, Feenberg presumably favors social relations with more equality of power and opportunity, and is willing to risk a fair amount of social stability in order to get them. Hammering out normative positions on specific instances of technological commodification (or decommodification) will demand a great deal of attention to detail. Nevertheless, one more general point can be made.

Though Feenberg allies himself with Albert Borgmann at a broad level, he faults Borgmann's writings on commodification and the device paradigm for offering no concrete potential for reform.[20] Larry Hickman, another Left-leaning philosopher of technology, is much harsher, characterizing Borgmann as nostalgic, romantic, and even reactionary.[21] Yet Borgmann's account of commodification and the device paradigm provides the philosophy we need for evaluating technological transformations of everyday life in normative terms. Devices put goods that we need at our disposal. All we need is the wherewithal to pay for them. As Borgmann himself argues, this in itself is not a bad thing, but as commodity relations expand, we find ourselves in a place where human purpose is reduced to getting the money we need to participate in commodity exchange. The work that our forefathers did with more recalcitrant things served immediate needs less efficiently, less effortlessly, but it also had the ability to situate them, to involve them, in a host of open-ended engagements with their families, their neighbors, and the natural world. This work thus became a source of identity, enjoyment, community, and meaning that is denied to people who see themselves as consumers, waiting for the weekend to come. In my reading, the regret that Borgmann associates with this loss is derived as much from his debt to Marx as through his adaptation of Heidegger.[22] Far from romantic nostalgia for times past, it is a particular reading of Marx's view that alienation of the worker from the work process is a positive social evil.

Yet it is true that aside from the cultivation of local community interests and certain kinds of aesthetic and recreational experience, Borgmann does not give us a program of positive action. There is, as Feenberg says, no program for the reform of technology. The theory of technological commodification provides a basis for addressing some of Borgmann's issues within the design process itself. It should be possible for scientists and engineers (and philosophers and activists) to consider a technology's potential for alienation of goods, altering exclusion costs or rival uses, and tendency toward standardization, and it should be possible in many cases to imagine alternatives that would affect each of these dimensions of our social relations in different ways. Such consideration would be a community project that would occur at the heart of research and development, and should surface criteria and concerns that scientists and engineers are capable of addressing with the skills that Feenberg associates with primary rationalization. I do not claim that the open-ended and engaging relations with other people and with the natural world will be the result in every case. No one, not even Borgmann, wants that. Yet attention to these dimensions of material technology can provide a framework for the kind of actor network that Feenberg wants to encourage, and will lead naturally to the secondary rationalizations that he advocates.

In conclusion, the theory of technological commodification articulates some dimensions of material technology that are, as Latour suggests, active in forming social relations. The theory is not itself normative, yet it does provide a conceptual framework that will prove useful for making normative and political arguments about when specific tools and techniques should be promoted, and when they should be resisted. It thus fills in some of the gaps between critical theory of technology and recent work in the social studies of science. It also provides some hooks on which to hang some longstanding Marxist criticisms of technology, as well as for bringing Borgmann and Feenberg more closely together. What has been offered, of course, is only a sketch of such a theory, yet if Feenberg finds this to be a friendly amendment to his own work, it is a sketch that points the way toward future work in the philosophy of technology.

Notes

1. Douglas C. North, *Institutions, Institutional Change, and Economic Performance* (New York: Cambridge University Press, 1990).
2. A. Allan Schmid, *Property, Power, and Public Choice: An Essay on Law and Economics* (New York: Praeger Press, 1987).
3. Andrew Feenberg, *Questioning Technology* (London and New York: Routledge, 1999); Andrew Feenberg, "Constructivism and Technology Critique: A Response to Critics," *Inquiry* (Summer 2000): 225–38.
4. Feenberg, *Questioning*; Andrew Feenberg, "From Essentialism to Constructivism: Philosophy of Technology at the Crossroads," in *Technology and the Good Life?*, ed. E. Higgs, A. Light, and D. Strong (Chicago: University of Chicago Press, 2000), 294–315.
5. Albert Borgmann, *Technology and the Character of Contemporary Life* (Chicago: University of Chicago Press, 1984).
6. Bruno Latour, *We Have Never Been Modern* (Cambridge: Harvard University Press, 1993).
7. Bruno Latour, *Science in Action: How to Follow Scientists and Engineers through Society* (Cambridge: Harvard University Press, 1987).
8. Don Ihde, *Bodies in Technology* (Minneapolis: University of Minnesota Press, 2002).
9. Garry Wills, *Inventing America: Jefferson's Declaration of Independence* (New York: Vintage Books, 1979).
10. Albert Borgmann, *Holding on to Reality: The Nature of Information at the Turn of the Millennium* (Chicago: University of Chicago Press, 1999). See especially chapter 9.
11. E. P. Thompson, *Customs in Common: Studies in Traditional and Popular Culture* (New York: The New Press, 1993), 185–283.
12. Bruno Latour, "A Door Must Be Either Open or Shut: A Little Philosophy of Techniques," in *Technology and the Politics of Knowledge*, ed. A. Feenberg and A. Hannay (Bloomington: Indiana University Press, 1995), 272–81.
13. Ruth Schwartz Cowan, *A Social History of American Technology* (Oxford: Oxford University Press, 1997), 78–80.
14. Deborah Fink, *Cutting into the Meatpacking Line: Workers and Change in the Rural Midwest* (Chapel Hill: University of North Carolina Press, 1998).
15. William J. Cronon, *Nature's Metropolis: Chicago and the Great West* (New York: W. W. Norton, 1991).
16. Andrew Feenberg, *Alternative Modernity* (Berkeley: University of California Press, 1996).

17. Paul Starr, *The Social Transformation of American Medicine* (New York: Basic Books, 1982).
18. Langdon Winner, *The Whale and the Reactor: A Search for Limits in the Age of High Technology* (Chicago: University of Chicago Press, 1986).
19. Feenberg, *Questioning*, 132.
20. Ibid., 189.
21. Larry Hickman, *Philosophical Tools for Technological Culture: Putting Pragmatism to Work* (Bloomington: Indiana University Press, 2001), 115–28.
22. Paul B. Thompson, "Farming as Focal Practice," in *Technology and the Good Life?*, ed. E. Higgs, A. Light, and D. Strong (Chicago: University of Chicago Press, 2000), 166–81.

CHAPTER NINE

ANDREW LIGHT

Democratic Technology, Population, and Environmental Change

T. C. Boyle's *A Friend of the Earth*,[1] tells the story of Tyrone Tierwater, a one time monkeywrencher and environmental avenger for "E. F.!" (Earth Forever!) whom we first meet in 2025 in his mid-seventies. Tierwater is now working for a character based on Michael Jackson, who in his semiretirement has employed the elder eco-warrior to help save some of the last remnants of a few dying species—warthogs, peccaries, hyenas, jackals, lions, and what is likely the last Patagonian fox. The not too distant environmental future painted by Boyle is a disaster. Global warming has finally caught up to us with a vengeance and even the secure shores of the United States are wracked by unmitigated cycles of flooding and drought seriously degrading most semblances of life as we know it.

To be sure, though, people, and some versions of progress, go on. While most affordable food and drink is limited to some combination of catfish and sake (very little else having survived decades of disastrous weather and a series of crop blights), and there are constant threats of new strains of life-threatening and highly contagious diseases, suburban development continues and new humans come into existence with the promise, at least in the developed world, of longer life spans. But Boyle does not give us anything like the overly optimistic views expressed by some conservative columnists who dismiss the need for global climate treaties; this is not an environmental future that is only felt with difficulty in the underdeveloped South requiring simpler economic readjustments for Americans without a substantial shift in lifestyle. The world has changed in this story for the worse and it is felt by everyone regardless of place or class.

While we follow what appears to be the last major transition in Tierwater's life, we learn the major events of his past through a series of interspersed chapters set in the late 1980s and early 1990s. In these parts of the book Boyle fictionalizes Earth First! into Earth Forever!, and satirizes what has so far counted as the heyday of radical environmentalism along with its fictional roots in the works of writers like Ed Abbey. The portrait, however, of the radical environmentalist as a somewhat young man is not unsympathetic. We are given ample descriptions of situations that explain how someone could go from living a mild-mannered suburban domestic existence to becoming a saboteur ready to create property damage to extractive industries and utilities whenever possible. Boyle also throws in some nice details for those with even a passing understanding of current environmental history, turning Tierwater's daughter into a version of Julia Butterflly Hill, acclaimed tree-sitter. But unlike Hill, Sierra Tierwater does not survive her multiyear stint in a giant redwood to write a best-selling book. After three years in the late 1990s, and many close calls with a forest company intent on dislodging her, she slips and falls while talking to her father on a cell phone. The tragedy leads to a depression resulting in Tierwater's most ambitious and certainly inhumane attempt at ecoterrorism described in the book.

It's called the "Cachuma Incident," a plan by Tierwater to poison the water supply of Santa Barbara with tetrodotoxin, "twelve hundred and fifty times more deadly than cyanide . . . mutated in the lab to adapt itself to fresh water."[2] While Tierwater describes it as his darkest moment, in the end he doesn't go through with the plan. But in describing why he doesn't do it we are offered a caricature of what many believe to be the classic dilemma of the committed environmentalist: "Though I'd steeled myself, though I seethed and hated and reminded myself that *to be a friend of the earth you have to be an enemy of the people*, . . . though this was the final solution and I the man chosen to administer it, when it came right down to it, I faltered. I did. Believe me. Give me that much at least" (my emphasis).[3]

Must one be an enemy of the people to be a friend of the earth? The annals of environmentalism are certainly littered with misanthropy in various forms and guises. But we do well to wonder whether it is necessary or helpful for the sake of the future of the planet to take such positions. By the end of this story we are given no easy answers. Tierwater, when confronted, admits that the sabotage he did manage didn't do much to prevent the disastrous state of the world at 2025. Would more monkeywrenching have made things different?

Would the poisoning of a major metropolitan water supply or any other effort to substantially decrease the human population have made any difference? Would even a quicker transition to more sustainable technologies—which we do see happening in this story out of sheer necessity—change the outcome? Boyle's answer seems to be no. He may be right. On that score we'll have to wait and see.

I start with this summary of *A Friend of the Earth* in part because I think it helps us to remember that all environmental writers engage in different forms of fiction, and by this I don't just mean novelists such as Boyle and Abbey. Like so many issues involving advocacy, the articulation of an environmentalism often involves a form of prediction where we try to warn our fellows of a possible future, painting a picture of how we can get there if it is good, or how to avoid it if it is bad. While among environmental scientists, sociologists, historians, economists, philosophers, and the like, it is not so much a literary imagination that is employed in this task; it still involves the creation of a fictional future, even where that future is based on empirical evidence or rational explanation. Because most environmental writers are fundamentally normative in their outlook—not content to just describe some bit of the world but also to try to improve it—much environmental writing, academic or not, takes on the character of a morality tale. While most of us are not as creative as Boyle is in *A Friend of the Earth*, we still tell stories about possible futures that either will redeem humans or damn us personally or collectively.

In chapter 3 of *Questioning Technology*[4] Andrew Feenberg takes up one of the central morality tales of the environmental movement and it comes surprisingly close to the question the Cachuma incident asks us to consider: Must one be an enemy of the people to be a friend of the earth? Specifically, Feenberg takes up one of the more famous incidents in the environmental literature where the veracity of this question was at issue, focusing on an analysis of Paul Ehrlich's conservative environmental politics and his debate with the progressive Barry Commoner. Ehrlich's political program, responding to what he saw as the devastating environmental effects of population growth, assumes, according to Feenberg, a stagnant model of technological evolution in the face of environmental problems and, in effect, an affirmative answer to the question raised above. Though certainly Ehrlich would never advocate something like the Cachuma Incident he did see numbers of people as the root of our environmental problems. In contrast, Feenberg presents Commoner's form of progressive environmental socialism as a precursor of the current claim increasingly found in environmental circles that certain forms of modernization are compatible with environmental sustainability. But while Feenberg lauds Commoner's broader views of the possibilities of technological reform he critiques Commoner's solutions to environmental problems. Commoner sought a basis for the resolution of environmental issues in labor solidarity, arguing that environmentalism was in the best interests of workers. The convergence of labor and environmental interests in new movements against globalization notwithstanding, an environmental labor movement has not emerged. Instead Feenberg imagines environmental issues more appropriately as an impetus for a democratic reform of technology. Environmental problems create pressures on consumers and citizens that can help to motivate a new technological politics. In turn, the ends of technological reform will be an environmentally sound technology.

I will first examine the differences suggested by Feenberg between the positions represented by Ehrlich and Commoner, tease out the role that the Ehrlich-Commoner debate played in the current configuration of environmental politics, and its relation to the misanthropic narrative I have illustrated above, and then examine the broader place of environmental politics in Feenberg's theory of the democratic reform of technology. At the end I will briefly offer my own views on how to resolve tensions between ecology, technology, and democratic theory. Feenberg embraces a process of democratic rationalization aimed at helping to reform technology around issues such as environmental concern. For me the prior question is not how environmental issues can motivate the reform of technology but instead, first, what sort of practices will motivate concern for the environment. Because I think there is a preference in Feenberg's analysis for democratic forms of environmental management, I'm hopeful that my own way of looking at this cluster of concepts is ultimately compatible with his.

Commoner versus Ehrlich

Feenberg claims that the debate in the 1970s between Paul Ehrlich and Barry Commoner was critical for environmentalists since it was one of the first illustrations of key disagreements that would later shape the various sides of the environmental movement over the possibility of the reform of technology. In assessing this claim I want to first look at the focus in this debate on population issues and then the broader implications of this exchange on the reform of technology, which is Feenberg's stronger claim in the chapter. Again, Ehrlich's position famously focused on population issues, leading him to a preference for at least a kind of "lifestyle politics," which may have put to the side substantive critiques of the political and economic systems that might perpetuate environmental problems. At worst, depending on one's views on such matters, his position may require the creation of coercive policies designed to achieve desired population goals. Commoner disagreed with this approach, seeing the road to solving environmental problems instead in a democratic process of reform of social, economic, and political structures. "It's the economy stupid!" we can imagine Commoner saying, in response to Ehrlich's campaigns for individual responsibility in limiting what he believed to be the key factor in environmental problems.

Feenberg sees a repetition of this debate in the later differences evinced between Earth First! (more like Ehrlich), and the environmental policies championed by the Oil, Chemical, and Atomic Workers union (more like Commoner), as well as the debate between the "Fundis" and "Realos" in the German Green Party.[5] He is certainly correct that calls for an undemocratic form of population control have shaped some poles in the movement. Feenberg cites for example the famous letter from "Miss Ann Thropy" to the Earth First! newspaper claiming that AIDS was a natural self-regulating mechanism on exploding populations in Africa.

But it is arguably the case that even a casual mapping of the terms of the debate between Ehrlich and Commoner on population control onto other important debates in the last few years does not quite work, or at least does not do justice to the complexities of these issues in the environmental movement. Mark Dowie provides a more thorough chronicle of the issue of population in his recent work on the history of environmentalism and confirms that indeed population concerns have been at the forefront of some splits in the movement.[6] Even so, Dowie also tells us that most mainstream organizations such as the National Audubon Society, the Natural Resources Defense Council, and the Wilderness Society do not endorse anything like a Malthusian approach to population growth.[7] Most of these groups would have to be characterized as embracing a position in between Ehrlich and Commoner entailing a combination of concern over both population pressures and economic and political structures that encourage overconsumption.

It is fair to say, however, that the Sierra Club has been plagued with this issue for much of its recent history. But as Dowie reports, debates in that organization have not been about population control per se (at least not on individual responsibility for birth rates) but rather about immigration.[8] Prominent activists in the Sierra Club have been heavily involved with the Federation for American Immigration Reform (FAIR), a group that tries to make anti-immigration sentiments "palatable by forming coalitions with environmental organizations."[9] Frank Orem, an active member of FAIR was also chair of the Sierra Club's national population committee and managed, with the help of others, to get the issue on the table by both the national organization and local chapters. This process led to a disastrous tainting of the Club as plagued by racists. Several highly visible referenda by the club over proposed endorsements of immigration restrictions in California have been very embarrassing for the membership. But at least as Feenberg represents the Ehrlich-Commoner debate, the issue between them was not national borders so much as international population policies. Still, we can read the Ehrlich-Commoner debate as presaging this divide.

Do such arguments go beyond the Sierra Club? Feenberg is correct that Earth First! and other radical environmentalists have said embarrassing things about population and immigration in the past, but it is a mistake to define these groups in terms of the occasional ramblings of some of their members. We should remember that Murray Bookchin has made an additional career for himself by exaggerating the importance of the Miss Ann Thropy articles and similar regressive claims by Dave Foreman, one time Earth First! leader, and Abbey. But his criticisms, even when they provide helpful cautions against misanthropy in environmental circles, sound like the construction of absurd straw men when they are focused on these examples.[10] When it got right down to it, the antihumanist elements of Earth First! lost their struggle for the soul of the organization (as they do in Boyle's narrative as well), which is now more oriented toward issues that appear to be lifted from the pages of Commoner and Bookchin himself. And when Bookchin finally got around to confronting Foreman face to face on these questions, rather than generalizing from a few lines here and there, he came across as more conciliatory toward Foreman than anything else.

When it gets right down to it, we shouldn't define the environmental movement by reference to the history of radical environmentalism. I made this mistake in my earlier work—seeing the debate between social ecologists and deep ecologists as definitive for understanding environmental politics—a mistake I have tried to overcome. Bookchin's organization in this divide, the Left Green Network, is now practically defunct, with remnants mainly at the Institute for Social Ecology. Earth First! made a lot of noise but wound up having less impact in the long run than promised, though the U.S. government played a role in its retrenchment. Even if population issues were at the center of debates between Earth First! and its opponents, this really doesn't amount to very much on the larger environmental scene. The more radical groups in that arena now, such as the Animal Liberation Front and the Earth Liberation Front, while probably seeing population as important, do not orient their direct action campaigns along these lines. Even though population has surely divided environmentalists in a variety of guises, by and large the range of mainstream environmental organizations see population control as a part, but only a part, of broader environmental initiatives, and certainly not the most important aspect of those initiatives in terms of their policy agendas. Much more important are the sorts of concerns that require large-scale economic changes such as support of a comprehensive agreement on greenhouse gas emissions.

Remember, though, that for Feenberg the issue is not just population as such but more accurately what the debate about population between Commoner and Ehrlich has to say about the possibility of the social control of technology. According to Feenberg:

> Fundamentalist environmentalism emphasizes control of growth because it can conceive of no change in the industrial order that would render it ecologically compatible. Technological determinism thus leads straight to a Malthusian position for which environmental and economic values must be traded off against each other. This is Ehrlich's position. Commoner's contrary views depend on a non-determinist philosophy of technology which admits the possibility of rational technical transformation. Only on this condition can growth and the environment be reconciled.[11]

Conceived as such, Ehrlich's views are closer to the old determinism of Jacques Ellul and Martin Heidegger and Commoner's position is closer to Feenberg's own views (absent the prediction of class solidarity on this issue). Feenberg goes on to argue that for Ehrlich technology was a "naturalized process"[12] producing one kind of thing without serious consideration given to alternative technologies that could ease the environmental burden of increased populations.

This mapping of a direct relationship between conceptions of the possible reform of technology and political responses to environmental problems is extremely helpful. Feenberg's contribution here goes farthest in helping us to understand the larger environmental implications of different philosophies of technology (a point I will return to briefly below). But as an indication of the terrain up for grabs in the environmental movement, this conceptualization needs some reworking as well.

If we take Feenberg's suggestion at the opening of this chapter that Ehrlich's position emerges later in the politics of Earth First! and the German Fundis in the Green Party, a problem arises for his later explanation of the relationship between technology and the politics of population control. By and large, Earth First!, the Fundis, and most of the groups who have endorsed some form of population control, are supporters of alternative technology not simply as a lifestyle choice but as a larger critique of industrial society. It's good to keep in mind that the FBI sting operation that finally hobbled Earth First! involved an alleged sabotage of a power generation plant. It would be hard to imagine Ehrlich endorsing that particular move. Neo-Luddism is often mistakenly lumped with the alternative technology movement in the pages of radical environmental journals, but the latter certainly endorses a view that substantive changes in technological culture are possible, which implies a kind of technological meliorism rather than a determinism.

If we want to find environmentalists who seem to accept Ehrlich's notion of a value-free technology, we have to look to the so called "third-wave" environmental groups that emerged in the Reagan era (such as the Environmental Defense Fund [EDF] and the National Wildlife Federation [NWF]), who have cozied up to business interests. Dowie argues that figures such as Jay Hair at the NWF, in embracing an American corporate ideology as a way of trying to get corporations to buy into environmentalism, have discouraged "the participation of their members and the public at large in the assessment of the environmental impacts of new technologies."[13] But the NWF does not embrace the lifestyle politics of Ehrlich and has more in common with Commoner in terms of its assessment that the root solutions to environmental problems must come out of the economic sector. While certainly their conclusion on how to use the economic system differs from Commoner's[14] the structure of the assessment is the same as Commoner's, and different from Ehrlich's.

It may be helpful here to employ a distinction I used some years ago to understand the differences between deep ecology and social ecology. If we are to look beyond the particular debates between Bookchin and deep ecology founder Arne Naess to see the larger picture of the philosophical positions at odds here, it helps to think of the deep ecology-social ecology debate as being a form of the larger disagreement between two groups I called "thin" environmental materialists (because their materialism does not necessarily imply an ontology) and environmental ontologists.[15] Environmental materialists argue that the appropriate human response to environmental problems must primarily involve an analysis of the causes of those problems in the organization of human society through the material conditions of capitalist (or state capitalist) economies, and the social and political systems that sustain those societies. Material conditions, such as who owns and controls the technological processes that are used to stimulate economic growth, expand markets, and consume natural resources, are for these thinkers the starting points for unpacking the complex web of environmental problems. In contrast, "environmental ontologists" see more potential in diagnosing environmental problems as primarily involving individual human attitudes toward nature. For environmental ontologists, social, political, and material problems are the symptom of a larger crisis involving the relation of the self with nature, not the root cause. The primary cause of environmental problems involves some assessment of our disconnection from nature, spiritual or otherwise. As such, the principal location of solutions to environmental problems for ontologists is to be found in changing the "consciousness" (for lack of a better term) of individual humans in relation to the nonhuman natural world.

Seen through this distinction, Commoner, Bookchin, and the EDF count as environmental materialists and Naess and Ehrlich count as environmental ontologists. Of course, there will be differences between theorists within these two camps. Ehrlich has nothing like the full-blown philosophical ontology we find in Naess, but still, as Feenberg points out, he did argue for something like a new set of "spiritual values" as a priority over economic and political reform, which resonates with my description of an environmental ontology. Unfortunately for Feenberg's argument, the view of technology as a naturalized process is found here across the divide between Ehrlich and the EDF. Counterfactually, we can easily imagine the EDF in more or less the same form as it is today, without this emerging commitment to the neutrality of technology. In fact, other third wave environmental groups, as Dowie reports, do not share this view on technology. While certainly there is more than one way to parse out different environmental views from each other, here we have an indication that when it comes to the role of technology in environmental questions we have something like three or four groupings and not two. Between the "naturalized" technology of Ehrlich and the nondeterminist substantivism of Commoner there are the bulk of environmentalists, who endorse a nondeterminist substantivism even if they are not always as democratically inclined as Commoner.

Finally, there is also a minor quibble that could be raised with using the differing views of Ehrlich and Commoner on technology as a benchmark for assessing the environmental consequences of future developments in the philosophy of technology.[16] In a response to a variety of critics, Albert Borgmann,[17] whom Feenberg groups with Ellul and Heidegger as a determinist and criticizes in a later chapter, clearly embraces the alternative environmental technologies endorsed by critics such as Gordon Brittan and Jesse Tatum. Borgmann's admiration, for example, for Brittan's particular embrace of wind power as an alternative energy technology is predicated on its affinity with Borgmann's notion of focal things (Brittan's windmills are better incorporated into the local landscape and culture than the massive windmill projects in California). Still, we have with Borgmann a nascent attempt to consider larger societal changes more consonant with Commoner's views than we might have predicted from Feenberg's analysis.

This is by no means a devastating critique for the role of the Ehrlich-Commoner debate in the picture that Feenberg paints. Feenberg admits that the debate has almost been forgotten today. And yet, viewed as a precursor to the later important debates in the environmental movement, it seems to me more rife with exceptions than with rules; less a guidebook to future debates in the movement than Feenberg might like. What we do have here is a good critique of Ehrlich, and nothing here takes away from that critique or suggests that the democratic reform of technology that Feenberg imagines is estranged from environmental concerns. Though I will not go into the issue at any length here, I think that if we were going to look for debates that have both oriented and divided the contemporary environmental movement we would do better to look at disputes involving issues of race and environmental justice.

As I suggested earlier, the population debate in the environmental movement has become more specifically an immigration debate. Anyone who might suspect that there is some connection between the emergence of immigration as a focal point for right-wing environmentalists and the rise of the environmental justice movement would be on the right track. Again, Mark Dowie, whose work chronicles the progressive backtracking of the environmental movement since the early 1970s, ends his book *Losing Ground* on a note of hope because he sees a rebirth in the environmental movement around issues of justice rather than lifestyle politics. Beyond the differing ideologies of animal rights activists, Earth First!ers, eco-socialists, liberal environmentalists, and neo-Luddites is an emerging recognition that environmentalists can and must rally around a common call for justice.[18] Following Dowie's lead, I would suggest that the principles of environmental justice adopted at the First National People of Color Environmental Leadership Summit in October 1991, and the Sierra Club centennial address by Executive Director Michael Fischer in May 1992—inviting "a friendly takeover of the Sierra Club by people of color"—to be the beginnings of the debates that will shape the movement into this new century.

Democracy as a Means but for What Ends?

I now turn to the more important question of the role of environmental issues in Feenberg's larger argument in *Questioning Technology*, concerning the importance of and possibility for the democratic reform of technology. Even a cursory glance at the book reveals that the reform of technology is the principal focus of Feenberg's work (both here and in his last two monographs), not recent environmental history. As I said at the beginning, Feenberg's basic idea on the relationship between the environment and technology is that environmental issues will help to press the necessity of the democratic reform of technology. In turn, a more democratically oriented technology will produce greener technologies, which will be better for the environment.[19] For Feenberg, the democratic rationalization of technology does not amount to a naïve attempt to throw open every technological question to non-experts. Democratization and participation can come in at more reasonable points in the process of design and production and can involve many representative steps, such as improving the influence of professional organizations in the process.

What bothers me at times though is the sense that Feenberg appears to see environmental reform as a black box in this process, or better yet, a green box. Environmental issues are potentially reduced to a motivation for democratic reform and a beneficial byproduct of that reform. Environmental sustainability is therefore possibly simplified in this process to a mere end. But there are many different ways that the environment could be renewed and many different ways that this end could be achieved. I would argue that what we have to think about is not just the democratic reform of technology as an end friendly for and responding to environmental pressures, but the form environmental renewal will take as an end as well. I have every reason to think that Feenberg would agree with such a claim.

Democracy and ecological concerns, as Feenberg points out in his analysis of Ehrlich, do not always go hand in hand. From Malthus to William Ophels, there have been plenty of environmental theorists who have concluded that a strong, often undemocratic, state is the only answer to environmental ills. What makes Feenberg's views different? Clearly it is the priority of democracy to his notion of the reform of technology. But how far does that priority get us? Let's restate the place of environmental issues in Feenberg's scheme. Environmental concerns enter Feenberg's theory on one end as part of the motivation for the reform of technology, and leave on the other end, as products of a reformed technological processes. But viewed thus, there is an unambiguous place for democratic concern only at the front of this process: democratic movements for the reform of technology motivated in part by environmental problems. What guarantees do we have, however, that the end products of a more democratic process will be more democratic environmental technologies? The only claim that could be made on the face of it is that the production of undemocratic green technologies out of a more participatory process of technological design would be inconsistent with the democratic foundation of the reformed technological process. A technological process rooted in democratic reform would be irrational to select undemocratic technologies as an output.

Perhaps it is too much to ask that the green technologies produced by this process be guaranteed to be democratic. After all, the virtue of democratic theory is that the process is more important than the product: we think a decision based on democratic principles is better simply because it came out of a democratic process even if its products are not as efficient as the product of a nondemocratic process. For example, a train schedule that is more responsive to the needs of local communities is better even if the trains do not always run exactly on time, as opposed to a fascist train schedule where the trains always run on time but only in locales designated by a tyrant. But in the case of environmental issues the priority of democracy is crucial not only as a valuable practice in itself but also perhaps as a guarantor of long-term environmental sustainability.

Before explaining this point further it will help to try to figure out what is meant by "democracy" in this discussion. Talking about "democratic technologies" and "democratic practices" can be very vague and confusing. Feenberg is clear, however, that when he talks about democratic rationalization or the democratic reform of technology he is not embracing a thick or strong account of democracy to the exclusion of all else. He does not maintain that all technological decisions need to be directly democratic, producing a prescribed set of values. Democracy for Feenberg seems to mean something simpler like "participation." A decision is democratic if it is open to participation by a range of agents in some process; it is undemocratic if it is closed to participation.

This understanding sounds right to me. Democracy has become a mystified term in some contemporary schools of moral and political philosophy. As an important case in point we often find overly prescribed notions of democracy in the technology literature. Richard Sclove's work is sometimes accused of this problem. A good example is also found in F. N. Laird's treatment of participatory processes in technological decision making. Laird describes democratic participation thusly:

> Truly democratic participation changes the outlooks and attitudes of participants. It makes people more aware of linkages between public and private interests, helps them to develop a sense of justice, and is a critical part of the process of developing a sense of community.[20]

But clearly this is too much to expect out of all democratic processes. Unambiguous cases of democratic practices, such as voting, are not made undemocratic if "the outlooks and attitudes" of the participants are not changed, nor when participants do not develop a sense of "justice" or "community" following participation in such practices. Voting is still democratic even if it is only merely participatory. Surely, democratic practices can lead to such outcomes beyond an occasion for participation, but they need not. This is not to say that I oppose thicker aspirations for our understanding of democracy. I have recently advanced versions of civic republicanism specifically for environmentalism.[21] We are mistaken, though, if we claim that all democratic practices must meet those aspirations in order to count as democratic practices.

If democratic technologies can be minimally understood as those that admit the possibility of public participation, what is the importance of participation to environmental processes? Feenberg identifies the importance of such practices in their potential to aid in a cultural transformation to one that would demand both more democratic and more sustainable forms of technology. If we assume that current unsustainable technologies of production and consumption will never change then Ehrlich's worries about growth in population become difficult to answer. More people will mean more intensive use of such technologies, more drains on natural resources, and hence more environmental devastation. While such a forecast is not true across the board—Feenberg gives a good example of how agricultural technologies of even the worst sort are often underutilized now—it will be true for many quasi-public goods that have discernable limits absent changes in regulation. Take water, for example. In some parts of the United States major water shortages have been the result of overutilization, which can be indexed to population growth. But such pressures do result in new technologies which in turn decrease the impact of population pressures on these problems.

The question for a more ambitious form of sustainability is how to generate changes in technology prior to the point of crisis in delivery of environmental services. Commoner, as Feenberg points out, did not sufficiently appreciate the importance of cultural transformation to spur changes in technology, lumping too much of such concern into an uncharitable assessment of lifestyle politics. But later Commoner did revise this view, and as Feenberg puts it, once he became "involved with movements against toxics and for recycling, he too [came] to recognize the importance of voluntarism in the environmental movement, not for the sake of self-imposed poverty, but as a source of cultural change."[22] This view certainly seems right to me. As I have argued at length elsewhere, public participation in environmental processes is important because it does help to create a "culture of nature," or at least a critical mass of people who have a stake in the landscapes around them and hence an interest in beneficial environmental reforms, including moves to more sustainable technological infrastructures.[23] When such interests are created, we need not resort, as Feenberg puts it, either to political or moral coercion in opposition to people's perceived interests or to "market incentives," to trick them into signing on to a range of environmental ends.[24] Instead, such practices create a rational scheme in which we may act without requiring such top-down stimulants.

What kinds of practices would get us this kind of cultural change, opening the door to a broader embrace of green democratic technologies? Feenberg seems to be happy with the outcome of technological reform residing in democratic control over environmental decision making—the setting of priorities, avenues for development of green technology, etc. The goal could however be democratic participation in on the ground environmental management as well, which might go further to continuing the kinds of cultural change needed to spur more technological evolution. One example of a practice that has this potential is public participation in a range of projects known as "restoration ecology."

Restoration ecology is the technological practice and science of restoring damaged ecosystems, most typically ecosystems that have been damaged by anthropogenic causes. Such projects can range from small-scale urban park reclamations to huge wetland mitigation projects. These projects admit to large-scale public participation. In the United States, for example, the cluster of restorations known collectively as the "Chicago Wilderness" project in the forest preserves surrounding the city, have attracted thousands of volunteers to help restore the native Oak Savannah ecosystems which have slowly become lost in the area.[25]

While there are many things that could be said about it, in general, restoration makes sense because on the whole it results in many advantages over mere preservation of ecosystems that have been substantially damaged by humans. But it must be remembered that this is a technological practice, very different in kind than acts of environmental preservation. Restoration requires more intense and active forms of environmental management than many other comparable environmental projects. As a technological practice, then, what are its goals? Clearly, the goals are to create something as close to the original as possible that is able to maintain and regenerate itself in the long run. In this sense, not only must the technology produce a sustainable environmental outcome, but the technology itself needs to be self-sustaining, and this will require a different kind of technology. But the goal should not just be to create new landscapes that can perpetuate themselves, but to create forms of producing new landscapes which will stimulate a stronger human cultural relationship with those places.

If we were to only focus on the fact that restoration is taking place and chalk that up as proof of the importance of the sort of environmental issues that will motivate a democratic reform of technology, we would be missing something quite important. What we would be missing is that the fact that landscapes are being restored is not nearly as important as the choices that we have to make about what to value in those restoration projects which produce those landscapes. Seen as a technological practice it is easy enough to focus on the end products of restorations only and then use this focus to endorse the best technologies that can produce the best restored landscapes. I would argue, however, that as a human technological practice that involves our cultural connection to nature, we need to think about what values are produced in that practice and how those values can best be made use of in the broader, long-term project of creating a sustainable society. The kinds of technological interventions we can make through restoration need to be gauged so that they admit to creating opportunities for public participation in these projects if we are to see them fulfill their full cultural potential.

When restorations are performed by volunteers then they are, in the way I have described the role of participation above, democratic. The value of this participation is not, however, justified in a vacuum. In the case of restoration, participatory practices get us better restorations because they create the sorts of cultural changes that Feenberg and I agree are a necessary prerequisite for the democratic reform of technology. Much of my work over the past several years has been dedicated to proving the proposition that restorations that are not produced by volunteers do not capture this democratic value and do not necessarily aid in the creation of communities committed to the protection of their local environments.

Stepping back from this example, then, what is needed as an outcome of the democratic reform of technology is not just more sustainable technology, in terms of their impacts on the environment, but technologies that are more amenable to strengthening our cultural ties with natural systems rather than separating us from them. For example, we don't just want to replace unsustainable agricultural practices with new ones that are lighter on the land and less likely to compound the environmental impacts of the addition of more users on a food delivery system. What is ideally needed are agricultural practices that help to remind people of their role in the food production process, and, if possible, give them a role in that process as stewards of their environment. In the case of restoration what is needed is the development of technologies that make it easier for people to help to maintain the ecosystems around them. This is not to say that we won't achieve sustainability until everyone raises their environmental consciousness to the point where they are all made aware of their individual impacts on the environment. I for one am more than happy with the kinds of structural changes in development patterns, for example, that garner energy savings without requiring people to choose to live differently than they would otherwise (such as encouraging more dense development in cities). But, where possible, we would do well to make it easier for people to insert themselves into the environmental processes around them and thus aim to stimulate the creation of environmental technologies that are more conducive for those kinds of relationships.

Given the cultural importance of encouraging actual participation in local environments, I would argue that we should not be asking only what we want environmentalism or environmental issues to do for us with respect to the reform of technology but instead we should first ask what sort of practices will produce a democratic environmentalism that in turn will shape the future direction of a democratic technology. Such a view would be consistent with Feenberg's idea of democratic rationalization, but it could be made more explicit. If environmentalism contains its own internal democratic end (here, the value of participation in local environments) then this end must be prioritized in the democratic reform of technology. There is a bit of chicken-and-egg reasoning here. Environmental ills will only serve as a motivation for technological reform if we feel a stake in environmental problems. Feeling a stake in environmental problems may require participation in nature in a direct sense. But democratic technological reform may be a key for encouraging participatory technologies and systems in environmental circles. There is no necessary paradox here, only a cautionary tale that democratic practices must become the glue that holds together reform of technology and renewal of the environment as these two sets of practices mutually effect each other.

Let's go back now to the beginning of this chapter. We do not need to be an enemy of the people in order to be a friend of the earth. Whatever it means to be a friend of the earth (and to be truthful, I've never quite understood the sentiment) requires instead that at least some of us engage in a cultural connection with the broader aims of environmentalism expanded to include the ends of humanism: building more just and sustainable communities of persons. At the end of *A Friend of the Earth*, it is Tierwater's personal connections to his ex-wife and now-deceased daughter that make it possible for him to start over and to see the promise of humanity even in the face of the great peril it has created. Though we are not offered an easy happy ending, at least we get a glimpse of an important morality tale that all of us should face: to be more morally responsible in relation to our effects on the planet we must form connections not with some abstract conception of nature but more concretely with each other as inheritors of the conditions we have created for ourselves. Perhaps this is the story we need to be telling each other more often.

Notes

1. T. C. Boyle, *A Friend of the Earth* (New York: Penguin Books, 2001).
2. Ibid., 277.
3. Ibid.
4. Andrew Feenberg, *Questioning Technology* (London: Routledge, 1999).
5. Ibid., 47.
6. Mark Dowie, *Losing Ground: American Environmentalism at the Close of the Twentieth Century* (Cambridge: MIT Press, 1996), 161.
7. Ibid., 162.
8. Ibid., 165–66.
9. Ibid., 163.
10. See Light, *Social Ecology after Bookchin*, especially introduction.
11. Feenberg, 47.
12. Ibid., 54.
13. Dowie, 116.
14. Though it should be noted that Feenberg points out that Commoner's more recent work accepts the market as one tool among others for achieving sustainability.

15. Andrew Light, ed., *Social Ecology after Bookchin* (New York: Guilford, 1998), chapter 11.
16. A point, as I said above, that I see as a more promising outcome of Feenberg's analysis here.
17. Albert Borgmann, "Reply to my Critics," in *Technology and the Good Life?*, ed. E. Higgs, A. Light and D. Strong (Chicago: University of Chicago Press, 2000), 341–70.
18. Dowie, 262.
19. Feenberg, 220.
20. F. N. Laird, "Participatory Analysis, Democracy, and Technological Decision-Making," *Science, Technology and Human Values* 18 (1993): 341–61, 345.
21. See Andrew Light, "Urban Ecological Citizenship," *Journal of Social Philosophy* (forthcoming).
22. Feenberg, 67.
23. See Light 2000.
24. Feenberg, 66.
25. W. K. Stevens, *Miracle Under the Oaks* (New York: Pocket Books, 1995).

CHAPTER TEN

EDWARD J. WOODHOUSE

Technological Malleability and the Social Reconstruction of Technologies

This is an appreciation, gentle critique, and modest extension of Andrew Feenberg's work on technological malleability and its implications for reconstructing technological civilization. I attempt to show that the "alternative modernity" project could be advanced by making three empirical-methodological moves to help target the philosophical analysis.[1]

First, delving deeper into substantive details of technological problems and potentials can help pose more precise scholarly puzzles; I illustrate this claim via the case of "green chemistry," the redesign of synthetic organic molecules to make them less dangerous. Second, rather than attempting directly and holistically to analyze entire political-economic systems (capitalism/ socialism), I propose that key elements of technological governance can be separated out for study; corporate executive discretion is one such element. Third, it seems to me that one important task of political philosophy is to propose new institutions and processes that can improve on extant ones, and I discuss two such political-economic innovations that could markedly strengthen technological innovators' regard for important public concerns. Before turning to these tasks, I begin by briefly reviewing some of Feenberg's main ideas concerning political-economic steering of technological innovation.

Feenberg on Technological Malleability

Feenberg's ideas about technological malleability and governance are complex and nuanced, so no brief summary can do justice to them. Nevertheless, for purposes of my analysis, the following components of the philosopher's thinking are an essential starting point.

First, he understands technology as inherently flexible and malleable, providing many possible alternative trajectories. Constructivist scholarship in Science and Technology Studies (STS) sometimes has been criticized for being apolitical, but Feenberg perceives that constructivist studies highlight the contingent nature of both scientific research and technological innovation. This can be interpreted as pointing "to the possibility of reshaping the technical world around us. Technophobic ideologies of the sort that emerged in the mass culture and politics of the 1960s underestimate the potential for reconstructing modern technology."[2]

Second, because technology has many unexplored potentialities (alternative paths that could have been, or now could be taken), no technological imperatives dictate the current social hierarchy (or other sociotechnical arrangements on which technology impinges). Rather, technology is a scene of social struggle, a "parliament of things."[3] Technologies thus are inherently political, embodying values and serving some people's values much better than others.

Third, technology is not just a means to certain ends, but is important in shaping the contours of contemporary life. Feenberg echoes Winner and Sclove: technology *is* legislation.[4]

Fourth, some social interests exercise far more influence than others in the negotiations occurring in the parliament of things. "Political democracy is largely overshadowed by the enormous power wielded by the masters of technical systems: corporate and military leaders, and professional associations of groups such as physicians and engineers. They have far more to do with control over patterns of urban growth, the design of dwellings and transportation systems, the selection of innovations, our experience as employees, patients, and consumers, than all the governmental institutions of our society put together."[5]

This is true for a variety of reasons including elite dominance of government, but Feenberg points especially to the lack of democracy in economic life. He asks, "Why has democracy not been extended to technically mediated domains of social life despite a century of struggles?"[6] Great gains could thereby be achieved, he believes.

If grossly disproportionate elite influence were somehow overcome, Feenberg asks, is it reasonable to suppose that "technological reform (could) be reconciled with prosperity when it places a variety of new limits on the economy?"[7] His answer is largely historical and practical: There have been many such limits in the past, such as requiring safer steam engines, and business has adapted just fine. Hence, he implies, new limits or design criteria can be added without necessarily undermining the material successes of capitalist economies.

The long-term result of such changes would be an "alternative modernity," characterized by technologies reshaped, perhaps radically, to promote rather different ends than those now prevailing. This would constitute "subversive rationalization"—"technological advances that can be made only in opposition to the dominant hegemony. It represents an alternative to both the ongoing celebration of technocracy triumphant and the gloomy Heideggerian counterclaim that "'Only God can save us' from techno-cultural disaster."[8]

Would subversive rationalization be tantamount to socialism? Feenberg believes it would, when socialism is understood less as a political movement than as a civilizational alternative involving "the realization of suppressed technical potentialities."[9]

> (S)ocialism is a coherent transformation in the very foundations of the social order. . . . (T)he path to this result . . . consists in three transitional processes: The socialization of the means of production, . . . an end to the vast economic, social, and political inequalities of class societies (and) a new pattern of technological progress yielding innovations that overcome the sharp division of mental and manual labor characteristic of capitalism.[10]

Some radical ecologists, communitarians, and other progressive scholars/activists may doubt that Feenberg's goals for governing technology logically entail these or other elements of "socialism," but I do not wish to focus on the terminology or even on what could be significant differences concerning the ultimate shape of a better system of technological governance. We are so far from understanding how to design a technological civilization that can escape what Winner calls "somnambulism," that progressive thinkers and activists of this generation have plenty to do in envisioning and working toward medium-term reforms. Whatever might ultimately divide us when the good society is closer to realization, I will assume for purposes of this analysis that Feenberg and I are aligned in seeking some new hybrid form of market and government that could govern technological innovation and use more wisely and fairly.

What modifications in existing institutions, or what new institutions, might enable technological decision making superior to that now characterizing contemporary political economies? Perhaps the most specific change Feenberg advocates is that of industrial democracy, which he considers the "underlying demand behind the idea of socialism." He believes "that unless democracy can be extended beyond its traditional bounds into the technically mediated domains of social life, its use-value will continue to decline, participation will wither, and the institutions we identify with a free society will gradually disappear."[11] Feenberg's interest in catalyzing democratic deliberation concerning technology to date has not led him to analyze systematically the shortcomings of contemporary governance institutions or to propose specific improvements. While I recognize that some division of scholarly labor is appropriate, I wonder whether political philosophers of technology might wish to spend a portion of their energies outlining hypothetical governance arrangements that could improve on existing practice. Toward that end, this chapter inquires into two possible political-economic procedures that potentially could bring about better steering of technological innovation. I turn to these issues after first echoing and reinforcing Feenberg's arguments about technological malleability.

Further Considerations Regarding Technological Malleability

Although Feenberg is emphatic and sometimes eloquent in arguing that technologies are malleable and that technological determinists therefore are misguided, his examples do not quite support the large edifice they are intended to help erect. His two main case studies (experimental AIDS pharmaceuticals and the French videotex experiment that made networked terminals widely available long before the Internet) seem to me inadequate to compel belief that technologies are highly malleable and could be directed quite differently than they now are. For that claim to be persuasive, one needs examples that go to the heart of contemporary life, not to its periphery. We need a repertoire of technological potentials demonstrating "the *contingency* of the existing technological system, the points at which it can be invested with new values and bent to new purposes."[12]

Coming closer to constituting such an example is a discussion in the opening pages of *Alternative Modernity* regarding the demise of the nuclear power industry.[13] What one thinks of as "nuclear power" was by no means the only way to convert atoms into usable energy. The gigantic light water reactors built primarily using GE boiling water and Westinghouse pressurized water designs resulted from historical contingencies, not from technical necessity. As early as 1956, some nuclear technologists recognized that it would be feasible to build very small reactors that could not suffer core melts. In the words of one proponent, the goal should be a reactor "so safe that it could be given to a bunch of high school children to play with, without any fear that they would get hurt."[14] This type of reactor would have "inherent safety," that is, safety "guaranteed by the laws of nature and not merely by the details of its engineering."[15] Although not used commercially thus far for producing electricity, technologists have investigated several designs for "inherently safe" reactors. These would not have done away with concerns over radioactive wastes, but would have been immune to core melt, most people's main worry. Tiny reactors not much bigger than those used by hospitals to create radioactive isotopes also would have been immune against nuclear proliferation or terrorist threat.

The small reactor designs were not pursued because of context-specific reasons having to do with social more than technical judgments. As the only utility executive to criticize the Atomic Energy Commission's hyper-optimistic cost estimates in the 1960s put it, there emerged "a bandwagon effect, with many utilities rushing ahead to order nuclear power plants, often on the basis of only nebulous analysis and frequently because of a desire to get started in the nuclear business."[16] Congress and the AEC likewise rushed ahead, helping induce utilities (e.g., by limiting their liability), in part because the nuclear nations were perceived to be in a "kilowatt race"; as one U.S. Senator put it, "If we are outdistanced by Russia in this race, it would be catastrophic. If we are outdistanced either by the United Kingdom or France, there would be tremendous economic tragedy."[17]

Military considerations also played a major role. Hyman Rickover in the late 1940s picked the light water reactor for the nuclear submarine program—because the reactor could be developed quickly and compactly for a submarine's confined space. Yet that program obviously was a Cold War response to the Soviet Union, and prior to commercialization there was no sustained deliberation concerning whether light water reactors would be especially safe, acceptable, or otherwise well suited to widespread commercial use. Because of the military experience, the pressurized water and boiling water reactors from General Electric and Westinghouse had built up an insurmountable lead over competing technologies. Thus, when the congressional Joint Committee on Atomic Energy pressured the Atomic Energy Commission to accelerate development of civilian nuclear power, these were the only reactors readily available.

This is a great example of Feenberg's confidence in technological malleability, but it does not make as good a case for democratization of technological choice, because Congress and the AEC drove the technology at least as much as did the corporate actors. Moreover, citizen groups used a combination of protest and legal action to choose what many considered a superior alternative in the end: not nuclear. Unlike gasoline-powered transport, the technology was rejected before full-scale deployment. Hence, the system worked, it could be said, so what is the philosopher complaining about? In some eyes, therefore, the nuclear example may not constitute an unimpeachable instance of malleable technologies badly shaped by social forces. The nuclear case also is unique in so many respects that one has to wonder whether the malleability thesis can be demonstrated for ordinary innovations, ones more a part of everyday life.

Green Chemistry

Exactly that sort of claim is now being made by advocates for "green chemistry," who aim to revolutionize the chemical industry by making synthetic organic chemicals environmentally sustainable.[18] Green chemistry advocates believe that chemicals and chemical production processes can be redesigned at a molecular level to render them inherently benign or at least radically less dangerous to people and to the environment. Given the centrality of chemicals in technological civilization, this case may demonstrate beyond doubt that technologies near the heart of contemporary life could be socially reconstructed in public-regarding ways.

The twentieth-century chemical industry proceeded by reacting two or more compounds to produce an intermediate product, which then was combined with other chemicals in a long series of "stoichiometric" reactions. Unwanted byproducts called hazardous wastes were produced at many of these steps, and had to be disposed of. The work was done in gigantic automated plants yielding many tons of final chemical products, which were released into ecosystems and human environments without knowledge of long-term effects, and without proceeding slowly enough to learn from experience prior to creation of unacceptable risks. As absurd as that "formula" seems in retrospect, chemical professionals of the past century used it with scarcely a whisper of doubt.

The green chemistry ideal would differ in every respect:[19]

- Design each chemical to be inherently benign, and/or to be quickly excreted from living organisms and quickly biodegraded in ecosystems;
- Create the chemical from a carbohydrate (sugar/starch/cellulose) or oleic (oily/fatty) feedstock rather than from depletable fossil fuels;
- Find alternative synthesis pathways requiring only a few steps, while creating little or no hazardous waste byproducts;
- Rely where possible on a catalyst, often biological, in a small-scale process using no solvents or benign ones;
- Create only small quantities of the new chemical for exhaustive toxicology and other testing;
- Scale up very gradually, revising precautions as experience reveals what can be done safely.

This chemical scenario appears potentially within the capacities of a revamped chemistry and chemical engineering.[20] Sometimes referred to as "sustainable chemistry" or "benign by design," green chemistry is billed as "promoting development of safer chemical production processes and products," via research "designed to reduce or eliminate the use or generation of hazardous materials associated with the design, manufacture, and use of chemical products."[21] In other words, what we have known as "chemistry" all these years turns out actually to have been a small subset of the overall possibilities for chemical design, and there is potential for a profound transformation in the methods, raw materials, byproducts, and end products of chemical synthesis.

Although research is well ahead of practice, promising industrial applications are beginning to appear. Manufacture of ibuprofen, the well-known painkiller sold as Advil™ and Motrin™, used to generate tons of hazardous wastes including cyanide and formaldehyde, whereas a new synthesis process eliminates these unwanted byproducts.[22] Likewise, instead of using 3.5 million pounds of ozone layer–depleting chlorofluorocarbons annually as the blowing agent for manufacture of polystyrene foam egg cartons and food containers, Dow Chemical now uses harmless carbon dioxide derived sustainably. Because of a new biodegradable chemical created for marine paints, ship maintenance facilities now can prevent growth of fouling organisms such as barnacles on ship hulls without using organotins that kill nontarget organisms and decrease reproductive viability of many aquatic species.

These small, mundane examples of greener chemistry appear to be harbingers of what is possible more generally—such as replacement of dangerous solvents such as benzene and toluene, of which millions of pounds now are used, by harmless solvents such as lactic acid (made from milk) and even water. It is important to acknowledge that green chemicals bear no necessary relation to the much broader social goals of Feenberg and other progressive observers of contemporary technology; a revised chemical industry may well constitute nothing more than "the American Way of Life equipped with emission controls."[23] My claim therefore is a very limited one: The story of brown versus green chemistry strongly and clearly reinforces the findings of STS scholars who have repeatedly demonstrated the contingent nature of technological innovation. Bhopals and Love Canals, endocrine disruption, ozone depletion, Multiple Chemical Sensitivity, and other chemical safety and environmental ills turned out the way they did not because chemical reality dictated it, but because university chemistry and chemical engineering departments, and their graduates in industry, cooperated with business executives to innovate in certain directions rather than in others.

The substantial malleability of the chemical universe also constitutes very strong support for Feenberg's ideas concerning the array of alternative technological trajectories that a more democratic decision-making process might evoke. As the next section attempts to demonstrate, understanding alternative technical trajectories such as green chemicals can help philosophers and social scientists target inquiries into questions that follow from the basic finding of malleability.

Inducing Business to Innovate Differently

What would it take to accelerate movement toward a greener chemicals industry (and toward making better use of other technologies' malleabilities)? In looking for specific institutional reforms, I have not found much guidance from philosophers of technology, or from other STS scholars, or among political scientists who study environmental and technological issues. Their conceptual analyses seem to me to focus at too high a level, and their empirical analyses are too close to existing practice; we need something in between.

Closer to that are the economists who recommend taxation (or subsidy) to raise (or lower) prices to better reflect the full social and environmental costs of products.[24] Several European governments have initiated environmental taxation, and it appears to be having some desirable effects in modifying production and purchasing patterns.[25] Instead of substituting centrally determined prices as did Soviet economic planners, and thereby running the risk of distorting entire industries' choices of factor inputs and production outputs, environmental taxes start from market-derived pricing and merely modify obvious shortcomings. That is a far more feasible task than setting prices from scratch, and it is a task too important to be left to economists and policy makers alone: philosophers of technology and other technological observers have important roles to play in thinking through the types of social costs/benefits that might be better captured in the pricing cues that help guide the buyers and sellers who collectively choose some important technological directions.

However, taxation and pricing is not my topic here. Instead, I want to probe a very different tack that may deserve consideration as a means of accelerating public-regarding technological innovation. My starting point is simply to ask: What gets in the way of wise shaping of inherently malleable technical potentials? And how might such barriers be reduced or circumvented? Whatever else might belong on the list of barriers, a significant one would be the possibility that chemical industry executives might use their discretion to choose not to switch rapidly to greener synthesis processes, perhaps partly because they try to keep costs/prices low so as to attract customers. Unless green chemical advocates can find ways to get around this obstacle, chemical greening is likely to proceed rather slowly. On the other hand, if we can learn something about circumventing or otherwise dealing with recalcitrant chemical executives, we may be able to generalize that lesson so as to improve technological governance more generally.

What might induce business executives to figure out superior ways of melding business profitability and technological innovation with social and environmental responsibility? Contrary to market ideology, competition and other market pressures by no means determine how business executives and their corporations behave. Of course innovators are guided by their understanding of what may sell profitably, but high-level executives exercise substantial discretion over at least four aspects of technological innovation:

- Executives exercise discretion over what new products to put on the market. Chemical executives and their R&D staffs developed chlorinated chemicals such as DDT and PCBs—partly to profitably dispose of the chlorine produced as an unwanted byproduct in the manufacture of chlor-alkali (used in many basic industrial processes).[26]
- Executives rather than consumers decide which technological innovations not to develop: For more than half a century after the invention and sale of nylon, scarcely any research was done on how to recycle the textile—and millions of pounds still end up in landfills every year.
- Corporate executives decide *how* products are put together technologically, with consumers at most choosing an outcome, not the process required to reach it. Commercial cleaning establishments and their industrial vendors have made perchloroethylene (PERC) a common, toxic constituent of urban air pollution. Probably not one customer in ten thousand knows how wool sweaters are dry cleaned, and comparative testing suggests that a majority actually prefer the smell and look of garments cleaned with safer solvents.[27]
- Corporate executives exercise many other kinds of discretion, including where to locate manufacturing plants and other large facilities; how often to update plant and equipment—as, for example, how soon to invest in new technologies for producing chemicals from corn and other biomass, and whether to invest in chemical synthesis pathways that make it feasible to recover and reuse chemical intermediates.

In sum, as has been true of most industries, a relative handful of men exercised disproportionate influence over what became the worldwide toxic chemical problem, with dioxins and other chlorinated compounds entering the tissues of most living creatures. Although buyers of course contributed by responding enthusiastically to the chemical products offered, toxic chemicals could not have become an environmental horror story without the initiatives taken using corporate executives' discretionary authority. What would it take to channel that discretion toward green chemistry?

Channeling Corporate Discretion

Observing the effects of corporate executive discretion in generations past, numerous social thinkers including Feenberg have found it a fatally flawed component of economic governance, one that should be replaced by industrial democracy and other components of a socialist political economy. Inasmuch as that dream has failed, at least for now, and corporation and market apparently will be around for the foreseeable future, it perhaps makes sense to probe for better ways of structuring corporate discretion. Is there some nontrivial way executives could be encouraged to innovate in more public-regarding ways?

One key is to recognize that business executives already work within certain constraints. To the extent that they act as agents of market buyers, as in pressing to keep prices low and variety high, they already are partly deferring to myriad others. Hence, the appropriate question is not whether corporate executives should share decision-making authority, but in what respects they should share which aspects of technological choices—and with whom?

Neither liberals nor conservatives, neither scholars nor practitioners have done a very good job of thinking through the matter. My few comments on the matter will do no more than show how open the problem is for future inquiry, but we can start by acknowledging that in the case of the chemical industry consumers are not going to rise up to demand green chemistry. Whether one wants industry to scale up more rapidly in making polyaspartic acid for plastics from corn, or whether one cares more about phasing out chlorinated chemicals, or whichever other aspect of chemical greening, it is hard to escape the fact that we are going to have to rely at least partly on industry executives who have chemists and chemical engineers on staff. The technologists and business managers inevitably will have an indispensable role to play in figuring out new formulas, investing in less hazardous production processes, marketing safer chemicals, and otherwise leading the way toward chemical greening.

But how to motivate CEOs and their high-level associates toward public-regarding innovation? No doubt decades or generations of research, public debate, and experimentation will be necessary to evolve a more intelligent form of corporate/market system, but a starting point is easy to discern merely by slightly restating the goal: we need some *mechanism for inducing* the most important business decision makers to exercise their discretion over technological innovation in ways that are better for environment and other public considerations. Socially improved innovation would require that corporate executives create a better ratio of public service to profitability. If it makes little sense for CEOs and their businesses to be rewarded for behaving in ways antithetical to what makes sense for humanity, then the trick is to find an incentive arrangement that will effectively revise the strange "contract" that now encourages socially perverse behaviors so long as they are profitable.[28]

Philosophers may wish to conceptually probe the set of possible incentive arrangements that could blend profitability with public service, but I merely want to show that it is not a null set. There definitely is at least one interesting and possibly effective option: change how high-level executives of the most important businesses are paid. Let them be hired and fired just as they are now: Company unprofitable, stock price falling? The board of directors may fire the old CEO, and bring in someone new. That makes as much sense as any other arrangement so far proposed for *choosing* top executives.

However, it is not at all clear that CEOs also should be *paid* in a manner set by the board (or by the executives themselves, in some cases). The combination of two incentives (job tenure and pay) pretty well guarantees that CEOs will prioritize corporate growth and profitability rather than public service when the goals conflict. Indeed, without incentives, there is little reason to expect them to try very hard to figure out how to make the goals more compatible. On reflection, then, it is hard to miss the possibility that CEOs might work harder for public goals if they could earn large salaries and bonuses only to the extent that their company improves in serving *public* purposes.

I recognize that the idea seems far-fetched on first inspection, but consider with me how it might be made to work. Suppose the maximum annual bonus for CEOs of the largest chemical companies were set at $10 million apiece. One or more of them could earn the maximum, and others could earn lesser amounts, based on how well each performs in competing to move their companies toward chemical greening. How much of a differential would be required to motivate the best possible performance? Ought the comparisons be done on an overall basis, or divided into categories such as consumer products, pesticides, and chemical intermediates? Should the scheme be extended beyond chemical greening to additional aspects of company performance? Such empirical and ideological questions would have to be answered by some combination of learning from experience and partisan negotiation, as would many other aspects of the revised incentive/accountability system.

A new approach to CEO pay obviously would require new oversight institutions, moreover, perhaps coupled with a complex "accounting" system for tracking chemical innovations and other aspects of corporate social responsibility. This might be overseen by an expert board equivalent to the U.S. Federal Reserve, or it might better be understood as a more partisan undertaking—one drawing, for example, on environmental interest groups as part of the governing unit. The experiment could even be coupled with evolving experiments using citizen juries and other new representative mechanisms.[29] However instantiated, any such major change running counter to long-established ways of doing business clearly would require controversial judgments.

Fortunately, many organizations already perform aspects of such accounting on which the new system could draw. The U.S. General Accounting Office monitors not only financial matters but also program accomplishments of governmental programs, and many of the methods could be carried over to the so-called private sector. The Environmental Defense Fund has an elaborate Web-based corporate scorecard based on the Toxic Release Inventory.[30] Analysts at mutual funds and at large securities firms continually monitor many aspects of the management of companies whose stocks they are recommending; and many other professionals study, rank, and assess myriad other aspects of corporate behavior.

No one is in a better position to figure out how to combine public purposes with private profitability than the CEO, and no one is in a better position to motivate other executives in the company (and the company's suppliers) to work for these public goals. Exactly where to start does not much matter, because it will take a century or more to learn our way into a new corporate incentive system; but the United States could begin with the Fortune 500 CEOs, or whatever other set of major corporations we can agree on, and expand/improve from there as experience reveals the shortcomings and strengths of the idea.

Improving What Gets Sold

The CEO-pay proposal responds primarily to weaknesses in *how* goods and services are produced. A second core weakness in technological steering concerns *what* is produced and marketed, and this problem may require a very different institutional approach. Just as one would not trust a patient to choose the right antibiotic for an infection, it is not clear how far to trust individual consumers to pick the right item in other situations. For example, only a minority of computer users have appropriately ergonomic desk/chair/keyboard trays; several billion extra dollars are spent annually on shampoos that product testers find work no better than the inexpensive brands; and obesity has become a worldwide disease afflicting 1.2 billion persons. The list of poor consumer choices could be extended almost indefinitely.

The persons directly involved are not the only ones who suffer, because many consumer choices also have broader public consequences. Pedestrians, bicyclists, passengers, and other drivers may be killed or maimed by cars with mediocre tires, and even in one-person accidents the costs of vehicle repair and medical care are shared via insurance. The cotton clothing now dominant worldwide is worse than most alternative textiles in terms of causing soil erosion, pesticide use, and in energy required for drying wet clothes—and probably no more than one shopper in a thousand knows it. Each used car maintained poorly and junked sooner than necessary translates into substantial energy and environmental consequences. Overall, it is harder to think of consumer choices that truly involve only the user than to list choices with fateful public consequences.

Thus, it would be easy to build a prima facie case against unrestrained consumer sovereignty.[31] What might be done? I believe it may be worth looking into institutional innovations that could achieve more competent ordering of goods and services from suppliers. Suppose for purposes of illustration that it makes sense to have less cotton and more polyester in textiles, and that the synthetic fiber should be derived from corn or another renewable crop rather than from petrochemicals? How might such a shift be induced? One modest modification of ordinary market system functioning might go a long way: instead of clothing manufacturers selling to wholesalers, who then sell to retailers, what if the intermediate wholesaling step were taken over by quasi-public organizations? Call them democratic wholesaling organizations until someone comes up with a better term.

If appropriately mandated, such organizations could be concerned with public issues such as energy and environment in ways that ordinary businesses cannot or do not. How might this system operate? Among many other challenges that would arise, consider three: (1) How to get manufacturers to go along with the shift? (2) How to get consumers to purchase the new mix of clothing? (3) How to pay for the changes?

First, consider the issue of production. How could the democratic wholesaling organization nudge farmers to grow less cotton and more corn, induce textile manufacturers to make more cloth from polyester and less from cotton, and induce clothing manufacturers likewise to play their part in the transition? Dead simple: gradually reduce wholesale orders for cotton clothing, while increasing orders for alternatives. For all its dreadful shortcomings, the great beauty of market exchange is that many participants in the system will react to changing conditions in search of sales.

But how to convince retailers to actually agree to sell less cotton clothing? There are two simple methods. First, raise the price charged to retailers for cotton clothing, and drop the price for fleece and other polyester. How much of a price change will induce how much of a shift in retailers' ordering is an empirical matter. Second, given that the democratic wholesalers have cut back on orders for cotton clothing, manufacturers will be unable to supply retailers' requests for quantities above that amount. Faced with a choice of stocking non-cotton clothing or none, few retailers will choose to leave store shelves empty so long as there is a reasonable prospect that customers will appear.

What about those customers, won't their tastes for cotton make them resistant to alternatives? Again, this is a question ultimately answerable only on the basis of experience. However, consumers' tastes are shaped in part by what they see others wearing, by advertising, by price signals, and by other contingencies. Over time, almost everyone modifies their clothing choices. To induce a more rapid than "natural" change, however, democratic wholesalers could join manufacturers and retailers in some of the less unsavory marketing tactics that now are deployed. At least as useful would be price inducements: by raising the price of cotton clothing and lowering the price of alternatives, many consumers will more or less automatically move toward the less expensive item.

Finally, how would the democratic wholesalers pay for all this public-regarding activity? Wholesalers presently are profit-making businesses, of course, and, if they get competent management, the democratic wholesalers likewise would enjoy a stream of revenue from retail customers sufficient to cover expenses and perhaps a bit more. Nevertheless, suppose for the sake of discussion that public-regarding products do cost more to manufacture. How might democratic wholesaling deal with the higher cost? An instance of this concerns the little-known controversy over chemical brightening agents in detergents, which consist of polymers that now are not biodegradable. Procter and Gamble and other major detergent manufacturers will not use the new water-soluble, biodegradable polymer that Rohm and Haas Chemical Company spent seven years developing, because it costs twice as much as the less environmentally friendly polymer now utilized.

That sounds like an insuperable obstacle—after all, how many people would purchase an environmentally friendly car that cost twice as much? In this case, however, the number is deceiving; there is so little brightening agent in laundry detergent that doubling the cost would bring a $4 box or bottle of laundry detergent to about $4.01.[32] Admittedly, that could add up to a million dollars in lost profits, some accountant has figured, a nontrivial sum, which is why we are having this discussion. Without some inducement, the soap industry is not likely to make the environmentally friendly move.

How might democratic wholesaling deal with the issue? Place orders only for detergents with biodegradable polymer as the brightening agent in place of the penny-cheaper, nonbiodegradable polymer now utilized. And pass the cost on to retailers, who presumably will pass the cost on to consumers, who will pay the higher price just as they now pay for other price increases. If no orders are placed with manufacturers who fail to use the biodegradable chemical whitener, then all detergent products on retail shelves will contain the environmentally friendlier ingredient. An alternative possibility is to squeeze manufacturer's profit margins; P&G is an enormously profitable enterprise, and it seems quite likely that executives there would prefer to cut profits rather than lose shelf space to a competitor. Democratic wholesaling could easily test just such a possibility, by gradually beginning to shift detergent business to Lever or another competitor willing to provide detergents with environmentally responsible ingredients at the same wholesale cost as the older, less environmentally friendly detergents.

The same process obviously can be applied to more important aspects of green chemistry.

Discussion

The constructivists are correct: technical feasibility alone does not explain the lines along which participants in innovation move—and don't move—toward technological stabilization and closure. The key methodological move that constructivists generally have not yet made, but should, is to pay equal attention to *undone* technoscience—that is, not only to study successful innovations such as bakelite, but to pay about equal attention to innovations that are not being developed, but arguably should be.

Feenberg also is correct: given that technologies are highly malleable, at least early on, and given that the particular form technologies take depends on social influences, the way is indeed open for a very different reportoire of technologies than those presently dominant—an "alternative modernity." Or, more accurately, *in principle* it seems *conceivable* to construct future innovations and to reconstruct existing technologies along many different lines, some of which could be less environmentally destructive, more egalitarian, more focused on quality of work life, more communitarian, or otherwise "better" than the present technological configuration.

This line of analysis regarding social reconstruction is part of a shift in Science and Technology Studies toward what might be termed neo- or post- or *re*constructivist scholarship.[33] While relying on the substantial contributions made by Bijker, Pinch, and others who developed the Social Construction of Technology (SCOT) tradition, as well as those of Actor-Network Theory (ANT) and other technology studies' innovations of the 1980s and 1990s, the reconstructivist project is attempting to bridge some of the gap between the descriptive/analytic and the more interpretive/politically engaged wings of the field. From a number of different directions, contributors are recognizing that there is no necessary conflict among the several types of intellectual agendas now present in the field, and seek to find ways that discipline-driven or disengagedly intellectual scholarship can benefit from the puzzles encountered by the more policy-oriented and activist scholars, and vice versa.[34] Such an engagement would stimulate greater reflexivity in the field as a whole, we think, leading to more carefully honed research foci—such as ones concerning undone science.

The central questions raised by reconstructivist scholars pertain to exactly the sort of topics that Feenberg's philosophy of technology is wrestling with: Stated most generally, how can better understanding of the actual workings of technological systems help open the way for more deliberate, wiser, and fairer reconstruction of particular technologies and of technological civilization more generally? What are the social obstacles to seeing malleability earlier and to making creative moves to circumvent these obstacles? What would it take to go beyond recognition that technologies are malleable and socially constructed, to actually have a reasonable chance of steering unfolding technological trajectories without the unacceptable sort of trial and error that accompanied the twentieth-century experience with chemicals?

As a democratic theorist, I share with Feenberg the goal of democratizing economic life, partly because a central canon of democratic theory is that citizens have a right to participate in decisions that affect them, partly because the potential intelligence of democracy can be actualized only by bringing into public choice the full variety of ideas, interests, and angles on problems that democracy at its best achieves.[35] However, I do not think that reiterating the goal of improved representation is enough; I have argued that philosophers and other students of technology need to think more deeply about institutional arrangements that would improve public accountability of the economic actors who exercise decision-making authority. Because it is far more difficult to envision new governance institutions than to criticize existing ones, I have argued that it makes sense to proceed somewhat modestly, in a more piecemeal fashion than Feenberg seems to endorse,[36] by dividing the very large task into smaller and more manageable components. A great many aspects of politics and economics and other social institutions come into play in producing technological change, so that I very much doubt whether we can understand and tackle the task holistically.

One approach would be to continue the line of investigation that led me toward the CEO pay and democratic wholesaling proposals: diagnose the bottlenecks and other obstacles standing in the way of public-regarding innovation, and then formulate revised social institutions and processes to circumvent, lower, or otherwise deal with the obstacles. Among many other candidate problems to which such a method might be applied are:

- Inappropriate expertise: University researchers may fail to develop certain socially beneficial technical potentials, technologists in industry may fail to apply emerging knowledge in industrial applications, and public-interest scientists in government, universities, and NGOs may fail to educate would-be advocates about the need for different technological R&D.[37]
- Shortcomings in interest articulation: NGOs may fail to take up the challenge, remaining content to publicize toxic releases as does the Environmental Defense Fund, overspending on fancy ships and cutting other staff as has befallen Greenpeace, or continuing to concentrate on charismatic megafauna as does the World Wildlife Fund.
- Mistargeted journalism: Mass media may focus excessively on symptoms of existing technologically mediated problems, failing to inform the public about technological malleability.
- Problems in consumer choice: Individuals, business buying agents, and government buying agents could continue purchasing older chemicals or other flawed products rather than demanding improved ones, perhaps partly because they are ill-informed and partly because they seek low prices.
- Inferior government policy: Officials may fail to catalyze public-regarding innovation trajectories.

Indeed, all of these obstacles have combined to slow down the shift toward greener chemicals. This was exemplified in recent deliberations in the U.S. House of Representatives on a tax credit to help dry cleaners purchase machinery to eliminate use of the toxic chemical PERC: staff received not a single inquiry from environmentalists, and the provision died in committee. Identifiable and potentially remediable elements of contemporary political economies lead to such results, and it seems to me that each of these obstacles to sociotechnical reconstruction deserve searching analysis by STS scholars, a task this chapter has barely begun.

In sum, I have argued for bringing into philosophical and political analysis of technology more information regarding specific technological problems and potentials. In particular, I have used the case of green chemistry to suggest that progressive thinking and action ought to include matters pertaining to the redesign of synthetic organic chemicals. It is time to move beyond cleaning up exhaust pipes and improving water treatment processes; chemists and chemical engineers can figure out how to better design virtually any production process so as to achieve chemical outcomes that fulfill human purposes. Given the centrality of chemicals to contemporary life, it is probably fair to assume that malleability applies in a great many technological arenas.

Recognizing malleability is just the beginning, of course, and we also need to figure out how to get the technologists and their corporate masters to move in the directions we seek. That implies examination of social institutions and processes, and redesign of them. I have argued that at least two steps would contribute to this task. One is to supplement high-flying critiques of capitalism with finer-grained analysis of key market and corporate processes that tend to misgovern technologies. Exactly what would have to change in order to move scientific inquiry, technological R&D, and product innovation in accord with positive potentials in green chemistry and in other technological arenas? Without denying many other obstacles to wiser, fairer technological governance, I have argued that corporate executive discretion and public-regarding wholesaling are perhaps the lynchpins. I do not see how technological innovation and utilization can be governed in significantly more public-spirited ways without somehow enrolling in the endeavor the most powerful decision makers in economic life.

Although no one well understands the full repertoire of possibilities, it is reasonably clear that discretionary choices made by business executives must better reconcile profit making with environmental preservation, quality of work life, and many other public needs. This might be done in a variety of ways, but the one that seems most promising to me is to alter the way top executives are paid. Remunerating the most influential CEOs on the basis of how well they achieve public purposes is the simplest, most direct, and most powerful method I can envision for inducing them to figure out how to better combine efficiency/profitability with improved outcomes for environment, workers, and other public concerns.

Consumption choices are by no means as powerful as is implied by the old notion of "consumer sovereignty," nevertheless approximately two-thirds of consumption is by households and it is difficult to imagine a civilization practicing wise consumption that does not involve myriad buyers making much more sensible purchasing decisions. Some extension of representative democratic procedures almost certainly would be necessary to assist with such a shift, and the most promising direction I now can see is that of intervening between manufacturer and retailer via democratic wholesaling.

In closing, another way to put the argument of this chapter is that we need to be even bolder than our leading observers of technology to date have been. However cogent the criticisms of philosophers of technology such as Feenberg, Winner, and Mitcham, and however compelling their visions of a more humane technological sphere, scholarship and activism alike could benefit from the three moves inculcated in this paper: More concrete probing into technical potentials, more precise analysis of defects in existing political-economic mechanisms, and more constructive efforts to specify improved sociotechnical governance mechanisms.

Notes

1. Andrew Feenberg, *Alternative Modernity: The Technical Turn in Philosophy and Social Theory* (Berkeley: University of California Press, 1995).
2. Ibid., ix.
3. Andrew Feenberg, "Subversive Rationalization: Technology, Power, and Democracy," in *Technology and the Politics of Knowledge*, ed. Andrew Feenberg and Alastair Hannay (Bloomington: Indiana University Press, 1995), 8.
4. Langdon Winner, *Autonomous Technology: Technics-out-of-control as a Theme in Political Thought* (Cambridge: MIT Press, 1977); Richard Sclove, *Democracy and Technology* (New York: Guilford, 1995).
5. Feenberg, "Subversive Rationalization," 3.
6. Ibid., 20.
7. Ibid., 13.
8. Ibid., 20.
9. Andrew Feenberg, *Transforming Technology: A Critical Theory Revisited* (New York: Oxford University Press, 2002), 147.
10. Ibid., 149.
11. Feenberg, "Subversive Rationalization," 4.
12. Feenberg, *Transforming Technology*, 189.
13. Feenberg, *Alternative Modernity*, pages 1–2, citing Joseph G. Morone and Edward J. Woodhouse, *The Demise of Nuclear Energy?: Lessons for Intelligent, Democratic Control of Technology* (New Haven: Yale University Press, 1989).
14. Edward Teller, quoted in Freeman Dyson, *Disturbing the Universe* (Princeton: Princeton University Press, 1979), 97.
15. Dyson, 97.
16. Philip Sporn, president of the American Electric Power Company, in a letter to the Joint Committee on Atomic Energy, December 28, 1967; reprinted in *U.S. Congress, Joint Committee on Atomic Energy, Nuclear Power Economics: 1962 Through 1967*, 90th Cong., 2d. sess. (February 1968), 2–22, quotation from p. 7.
17. Statement by Sen. John O. Pastore (D-R.I.), in *Joint Committee on Atomic Energy, Accelerating the Civilian Reactor Program;* quoted in Frank G. Dawson, *Nuclear Power: Development and Management of a Technology* (Seattle: University of Washington Press, 1976), 105.
18. This section draws in part on E. J. Woodhouse, "The Social Reconstruction of a Technoscience?: The Greening of Chemistry," in *Synthetic Planet: Chemical Politics and the Hazards of Modern Life*, ed. Monica Casper (New York: Routledge, 2003).
19. The term "green chemistry" as normally used at present does not actually imply adherence to all the ideal attributes I have put together into a single list.
20. A quick introduction to the subject is found in the field's first text, Paul T. Anastas and John C. Warner, *Green Chemistry: Theory and Practice* (New York: Oxford University Press, 1998).
21. www.grc.uri.edu/programs/2000/green.htm.
22. *The Presidential green chemistry challenge awards program [microform]: summary of 1996 award entries and recipients* (Washington, DC: U.S. Environmental Protection Agency, Office of Pollution Prevention and Toxics, 1996).

23. Feenberg, *Questioning Technology*, 65.
24. See, for example, Jonathan M. Harris. *Environmental and Natural Resource Economics* (New York: Houghton Mifflin, 2002).
25. "The Green Tax Shift," accessed May 30, 2003, http://www.progress.org/banneker/shift.html; Center for a Sustainable Economy, "Environmental Tax Reform: The European Experience," July 12, 2001, accessed May 30, 2003, <http://www.sustainableeconomy.org/eurosurvey.htm>.
26. On this and many other facets of the history, politics, and science of chlorinated chemicals, see Joe Thornton, *Pandora's Poison: Chlorine, Health, and a New Environmental Strategy* (Cambridge: MIT Press, 2000).
27. Peter Sinsheimer et al., "Commercialization of Professional Wet Cleaning: An Evaluation of the Opportunities and Factors Involved in Switching to a Pollution Prevention Technology in the Garment Care Industry," Final Report, October 28, 2002, Pollution Prevention Education and Research Center, Occidental College, accessed June 6, 2003 at <http://departments.oxy.edu/uepi/pperc/resources/Finial%20Report%202.0.pdf.>.
28. Of course, I do not mean to deny that some economists and many conservative thinkers would agree with Milton Friedman that a corporation's only responsibility is to the shareholders. And their argument would not be entirely specious: producing shareholder value requires creating goods and services that will sell, and, buyers may well perceive themselves better off as a result of the purchase—or they would not have parted with the money necessary to obtain it. There are many good reasons not to fully accept this logic, but it is not absurd.
29. Patrick Hamlett, "Technology Theory and Deliberative Democracy," *Science, Technology, & Human Values* 28 (2003): 112–40.
30. The EDF toxics scorecard can be accessed at www.scorecard.org.
31. On the other hand, markets coordinate complex, far-flung economic transactions far more deftly than any known alternative, as explained in Charles E. Lindblom, *The Market System: What It Is, How It Works, and What to Make of It* (New Haven: Yale University Press, 2001). So the challenge facing progressives is to take advantage of market strengths while correcting market weaknesses, a task that few have faced up to very squarely.
32. Estimate by Rohm and Haas chemist, offered at the NAS conference on Green Chemistry and Green Chemical Engineering, Washington, DC, June 1998.
33. See Edward Woodhouse, David Hess, Steve Breyman, and Brian Martin, "Science Studies and Activism: Possibilities and Problems for Reconstructivist Agendas," *Social Studies of Science* 32 (April 2002): 297–319.
34. See, for example, Patrick Hamlett, "Technology Theory and Deliberative Democracy," *Science, Technology, & Human Values* 28 (2003): 112–40, who calls for social constructivism to "broaden its connections with larger, normative questions" (quote from p., 135).
35. On the potential intelligence of democracy, see Charles E. Lindblom and Edward J. Woodhouse, *The Policy-Making Process*, third edition (Englewood Cliffs, NJ: Prentice-Hall, 1993).
36. The classic work on piecemeal or incremental analysis is David Braybrooke and Charles E. Lindblom, *A Strategy of Decision* (New York: The Free Press, 1963).

37. On appropriate and inappropriate expertise, see Edward J. Woodhouse and Dean A. Nieusma, "Democratic Expertise: Integrating Knowledge, Power, and Participation," pp. 73–96 in *Knowledge, Power, and Participation in Environmental Policy Analysis*, ed. Matthijs Hisschemöller et al., Policy Studies Review Annual vol. 12 (New Brunswick, NJ: Transaction Publishers, 2001).

ANDREW FEENBERG

Replies to Critics

An Autobiographical Note

I want to thank my critics for the intelligence and sympathy with which they have discussed my work. They have helped me immeasurably to see the weak points and especially the lacunae in my argument. I will attempt here to respond to those parts of their discussion I am able to weave into a coherent chapter at this time.[1] Many other ideas in these chapters will stay with me and provoke further thoughts long after this book has gone to press. Let me begin now with a brief note on the background against which my work is set. I will then turn to a sketch of the theory and the specific criticisms and suggestions found in these pages.

In the mid-1960s when I arrived in La Jolla to study with Marcuse, the New Left was still a very small phenomenon. Most criticism of American society was cultural criticism, much influenced by the Beats and Zen, or by the cultural elitism of intellectuals appalled at comic books, television, and rock music. In this situation political revolution seemed even more implausible than it does today. Marcuse's theory reflected the failure of the Marxian notion that capitalism was fraught with internal tensions between workers and capitalists.

Nevertheless, there were tremendous tensions of another type in this very conformist society, and they finally exploded the cozy world of the 1950s. The question was, what did those tensions mean, from what sources did they emerge, what was their object and destiny? Many of us who were living those tensions had a different answer from Marcuse. He followed his own teacher, Heidegger, more than he admitted to himself and to us in holding that the one-dimensional technological universe of advanced industrial society was a closed or nearly closed system in which opposition was impossible or nearly impossible.[2] When he began to see evidence of widespread opposition, he theorized it as a new cultural dispensation—Heidegger would have said "revealing." As we read Marcuse then, it seemed that the source of this revealing lay in the instincts and was thus external to the one-dimensional society.

I have recently completed a book on Heidegger and Marcuse in which I develop a more sympathetic reading of Marcuse's theory of revolution, but at the time many of us thought that this theory was too negative and unhistorical.[3] As we created movements against the war in Vietnam and engaged in student protest, we felt that our actions reflected internal tensions within one-dimensionality itself and not an external intervention from a transcendent source. But how to explain this without relinquishing Marcuse's insight into the integration of society and returning to the discredited concept of proletarian revolution?

I recall writing a long essay for Marcuse in 1966 called "Beyond One-Dimensionality" in which I tried to show how a one-dimensional society could yield up a new dialectic. Although my current work on technology is quite different from this early effort, the pattern is similar. I am still looking for a way of identifying and explaining internal tensions in a quasi-one-dimensional technological universe. In this my work departs not only from the first generation of the Frankfurt School but also from the contemporary consensus in political philosophy.

Under the influence of both Rawls and Habermas, political philosophy abstracts systematically from technology and so overlooks the dystopian potential of advanced society. It regards the technical sphere as a neutral background against which individuals and groups pursue personal and political goals. These goals are usually seen as more or less rationally justifiable ideas about rights, the good life, and so on. In this conception, technology simply cancels out as a constant and politics is a matter of opinion. As a philosopher of technology, I reject this view. What it means to be human is decided not just in our beliefs but in large part in the shape of our tools. And to the extent that we are able to plan and manage technical development through various public processes and private choices, we have some control over our own humanity.

After the 1960s the surviving sources of Left initiative were such non-Marxist movements as feminism and environmentalism. The society that had been the object of global condemnation in the 1960s was now challenged in concrete and specific ways. Industrial pollution, childbirth practices, experimental treatment of AIDS, all were contested by these new social movements in terms of the consequences of technical designs for human lives, health, and dignity. Comparable issues appeared in the labor movement around the deskilling of production technology.

My first book on philosophy of technology, *Critical Theory of Technology*, followed Marcuse in arguing that "technology is ideology."[4] But I emphasized an aspect of Marcuse's position that has not been widely noted, namely, his claim that the politics of technology depends on contingent features of technical design determined by a civilizational project, and are not due to the "essence" of technology in Heidegger's sense. This approach suggests that different designs might support a more democratic society based on democratic self-organization in the technical sphere itself.

The argument of this book departed from Marcuse's in an important respect. We are much more skeptical today of references to "nature" than in the period when Critical Theory was formulated. To avoid the naturalism and essentialism with which Marcuse is often (unfairly) charged I attempted to construct a non-ontological critique of technology that would nevertheless conserve the critical force of Marcuse's ontological critique. I concluded that wherever social relations are mediated by modern technology, it should be possible to introduce more democratic controls and to redesign the technology to accommodate greater inputs of skill and initiative.

These abstract arguments were a reflection not only of my reading of Marcuse but also of an extraordinary opportunity I enjoyed to participate in another kind of revolution, the computer revolution. In 1982 I was asked to help create the first program of on-line education.[5] At the Western Behavioral Sciences Institute, we employed a computer network to communicate with students for an extended program of study long before the Internet was opened to the public. This involvement brought me directly into contact with an emerging technical field and obliged me to learn its rudiments. I witnessed the role of human action in orienting the development of technology. Democratic aspirations for technology made sense in this context as we reinvented the computer to serve the humane purposes of education. Later, when the automation of higher education on the Internet was proposed, I saw my own theory of the ambivalence of technology exemplified in practice. I analyze this example in *Transforming Technology*.

Sociology of technology was undergoing a revolution of its own in the 1980s with the emergence of the contending schools of social constructivism and actor network theory in England and France. I was aware of these debates and learned much from them, but was dissatisfied with the refusal of both schools of thought to engage with the larger issues of modernity raised by the Frankfurt School. Yet the new sociology of technology did offer a fruitful methodology and powerful arguments against technological determinism that could be employed to support the idea of democratic change in the technical sphere. My approach is informed by contemporary technology studies and so achieves a level of concreteness Marcuse did not reach in his work. Nevertheless, I believe it can be loosely attached to the tradition to which Marcuse contributed. I therefore call it "critical theory of technology."

I now trace the tensions in the "one-dimensional" society to the difference in the way in which the world is experienced by those who administer it and those who are subject to it. Many years of struggles over technology in fields as diverse as medicine and environmentalism have shown that this difference is politically significant and prevents the closure Marcuse feared while also confounding the outdated schemas of traditional political philosophy.

Had I only known, Marcuse himself provided the basis for this analysis in one of his early essays, written under the influence of and in reaction to Heidegger. Marcuse asks: "Is the world 'the same' even for all forms of *Dasein* present within a concrete historical situation? Obviously not. It is not only that the world of significance varies among particular contemporary cultural regions and groups, but also that, within any one of these, abysses of meaning may open up between different worlds Precisely in the most existentially essential behavior, no understanding exists between the world of the high-capitalist bourgeois and that of the small farmer or proletarian. Here the examination is forced to confront the question of the material constitution of historicity, a breakthrough that Heidegger neither achieves nor even gestures toward."[6]

> The theory I am developing carries out the program implicit in these remarks. I conceive of technical arrangements as instituting a "world" in something like Heidegger's sense, a framework within which practices are generated and perceptions ordered. Different worlds, flowing from different technical arrangements and different positions within them, privilege some aspects of the human being and marginalize others. Goals flow from the nature and limits of worlds and not from arbitrary opinions. The clash between different worlds is inevitable in a society based on technological domination.

These arguments were developed in *Alternative Modernity* and *Questioning Technology*. In these books I moved from a post-Marxist to what I call a "critical constructivist" position and attempted to develop a more empirical orientation toward the study of technology.[7] The new edition of *Critical Theory of Technology*, entitled *Transforming Technology*, brings the earlier exposition of the theory into line with this later position. The three books now present different aspects of the same basic theory. It is this body of work to which my critics respond here.[8]

A Sketch of the Theory

Technology and Finitude

I want to begin this preliminary sketch of my theory by recalling the most basic problematic of my work. Radical critics of technology, from Mumford and Marcuse on down to the present, generally agree that the rise of technocratic power East and West has overshadowed class struggle. I too argue that the central issue for politics today is the prevalence of technocratic administration and the threat it poses to the exercise of human agency. This leads me to emphasize the essentially hierarchical nature of technical action, the asymmetrical relation between actor and object which, when it overtakes large swaths of human relations, tends to create a dystopian system.

This argument is meant to draw out the most basic implications of the Frankfurt School's critique of technology. I formulate this position in systems theoretic terms, distinguishing the situation of a finite actor from a hypothetical infinite actor capable of a "do from nowhere."[9] The latter can act on its object without reciprocity. God creates the world without suffering any recoil, side effects, or blowback. This is the ultimate practical hierarchy establishing a one-way relation between actor and object. But we are not gods. Human beings can only act on a system to which they themselves belong. As a consequence, every one of their interventions returns to them in some form as a feedback from their objects. This is obvious in everyday communication where anger usually evokes anger, kindness kindness, and so on.

Technical action represents a partial escape from the human condition. We call an action "technical" when the impact on the object is out of all proportion to the return feedback affecting the actor. We hurtle two tons of metal down the freeway while sitting in comfort listening to Mozart or the Beatles. This typical instance of technical action is purposely framed here to dramatize the independence of actor from object. In the larger scheme of things, the driver on the freeway may be at peace in his car but the city he inhabits with millions of other drivers is his life environment and it is shaped by the automobile into a type of place which has major impacts on him. So the technical subject does not escape from the logic of finitude after all. But the reciprocity of finite action is dissipated or deferred in such a way as to create the space of a necessary illusion of transcendence.

Heidegger and Marcuse understand this illusion as the structure of modern experience. According to Heidegger's history of being, the modern "revealing" is biased by a tendency to take every object as a potential raw material for technical action. Objects enter our experience only in so far as we notice their usefulness in the technological system. Release from this form of experience may come from a new mode of revealing but Heidegger has no idea how revealings come and go.

Like Marcuse, I relate the technological revealing not to the history of being, but to the consequences of persisting divisions between classes and between rulers and ruled in technically mediated institutions of all types. Technology can be and is configured in such a way as to reproduce the rule of the few over the many. This is a possibility inscribed in the very structure of technical action, which establishes a one-way direction of cause and effect.

Technology is a two-sided phenomenon: on the one hand the operator, on the other the object. Where both operator and object are human beings, technical action is an exercise of power. Where, further, society is organized around technology, technological power is the principal form of power in the society. It is realized through designs that narrow the range of interests and concerns that can be represented by the normal functioning of the technology and the institutions which depend on it. This narrowing distorts the structure of experience and causes human suffering and damage to the natural environment.

The exercise of technical power evokes resistances of a new type immanent to the one-dimensional technical system. Those excluded from the design process eventually notice the undesirable consequences of technologies and protest. Opening up technology to a wider range of interests and concerns could lead to its redesign for greater compatibility with the human and natural limits on technical action. A democratic transformation from below can shorten the feedback loops from damaged human lives and nature and guide a radical reform of the technical sphere.

Operational Autonomy

For many of critics of technological society, Marx is now irrelevant, an outdated critic of capitalist economics. I disagree. I believe Marx had important insights for philosophy of technology. He focused so exclusively on production because production was the principal domain of application of technology in his time. With the penetration of technical mediation into every sphere of social life, the contradictions and potentials he identified in technology follow as well. In my work I attempt to bring Marx's theory to bear on the general theme of technocratic power.

In Marx the capitalist is ultimately distinguished not so much by ownership of wealth as by control of the conditions of labor. The owner of a factory has not merely an economic interest in what goes on within it, but also a technical interest. By reorganizing the work process, he can increase production and profits. Control of the work process, in turn, leads to new ideas for machinery and the mechanization of industry follows in short order. This leads over time to the invention of a specific type of machinery which deskills workers and requires management. Management acts technically on persons, extending the hierarchy of technical subject and object into human relations in pursuit of efficiency. Eventually, professional managers represent and in some sense replace owners in control of the new industrial organizations. This is what Marx qualifies as the impersonal domination inherent in capitalism in contradistinction to the personal domination of earlier social formations. It is a domination embodied in the design of tools and the organization of production. In a final stage, which Marx did not anticipate, techniques of management and organization and types of technology first applied to the private sector are exported to the public sector where they influence fields such as government administration, medicine, and education. The whole life environment of society comes under the rule of technique.

The entire development of modern societies is thus marked by the paradigm of unqualified control over the labor process on which capitalist industrialism rests. It is this control that orients technical development toward disempowering workers and the massification of the public. I call this control "operational autonomy," the freedom of the owner or his representative to make independent decisions about how to carry on the business of the organization, regardless of the views or interests of subordinate actors and the surrounding community. The operational autonomy of management and administration positions them in a technical relation to the world, safe from the consequences of their own actions. In addition, it enables them to reproduce the conditions of their own supremacy at each iteration of the technologies they command. Technocracy is an extension of such a system to society as a whole in response to the spread of technology and management to every sector of social life. Technocracy armors itself against public pressures, sacrifices values, and ignores needs incompatible with its own reproduction and the perpetuation of its technical traditions.

The technocratic tendency of modern societies represents one possible path of development, a path that is peculiarly truncated by the demands of power. Like Dewey, I believe technology has other beneficial potentials that are suppressed under the capitalism and state socialism that could emerge along a different developmental path. In subjecting human beings to technical control at the expense of traditional modes of life while sharply restricting participation in design, technocracy perpetuates elite power structures inherited from the past in technically rational forms. In the process it mutilates not just human beings and nature, but technology as well. A different power structure would innovate a different technology with different consequences.

This argument goes to the heart of one of Doppelt's objections. It is in the context of technocracy that agency appears as a central democratic value not just for excluded minorities, as Doppelt claims, but for everyone.

Is this just a long detour back to the notion of the neutrality of technology? I do not believe so. Neutrality generally refers to the indifference of a specific means to the range of possible ends it can serve. If we assume that technology as we know it today is indifferent with respect to human ends in general, then indeed we have neutralized it and placed it beyond possible controversy. Alternatively, it might be argued that technology as such is neutral with respect to all the ends that can be technically served. However, I do not hold either position. There is no such thing as technology as such. Today we employ this specific technology with limitations that are due not only to the state of our knowledge but also to the power structures that bias this knowledge and its applications. This really existing contemporary technology is not neutral but favors specific ends and obstructs others.

A fuller realization of technology is possible and necessary. We are more and more frequently alerted to this necessity by the threatening side effects of technological advance. Technology "bites back," as Edward Tenner reminds us, with fearful consequence as the deferred feedback loops that join technical subject and object become more obtrusive.[10] The very success of our technology in modifying nature ensures that these loops will grow shorter as we disturb nature more violently in attempting to control it. In a society such as ours, which is completely organized around technology, the threat to survival is clear.

Resistance

What can be done to reverse the tide? Several chapters of this book propose alternatives to or amplifications of my conclusions. I welcome these contributions, but here I would like to explain my own argument, which can ground the various political and institutional suggestions of my critics at a more basic philosophical level.

The democratization of technology requires in the first instance shattering the illusion of transcendence by revealing the feedback loops to the technical actor. The spread of knowledge by itself is not enough to accomplish this. For knowledge to be taken seriously, the range of interests represented by the actor must be enlarged so as to make it more difficult to offload feedback from the object onto disempowered groups. But only a democratically constituted alliance of actors, embracing those very groups, is sufficiently exposed to the consequences of its own actions to resist harmful projects and designs at the outset. Such a broadly constituted democratic technical alliance would take into account destructive effects of technology on the natural environment as well as on human beings.

Democratic movements in the technical sphere aim to constitute such alliances. For my account of these democratic resistances, I rely on the work of Michel de Certeau.[11] De Certeau offers an interesting interpretation of Foucault's theory of power, which can be applied to highlight the two-sided nature of technology. He distinguishes between the strategies of groups with an institutional base from which to exercise power and the tactics of those subject to that power and who, lacking a base for acting continuously and legitimately, maneuver and improvise micropolitical resistances. Note that de Certeau does not personalize power as a possession of individuals but articulates the Foucauldian correlation of power and resistance. This works remarkably well as a way of thinking about immanent tensions within technically mediated organizations, not surprisingly, given Foucault's concern with institutions based on scientific-technical "regimes of truth."

Technological systems impose technical management on human beings. Some manage, others are managed. These two positions correspond to de Certeau's strategic and tactical standpoints. The world appears quite differently from these two positions. The strategic standpoint privileges considerations of control and efficiency and looks for affordances, precisely what Heidegger criticizes in technology. My most basic complaint about Heidegger is that he himself adopts unthinkingly the strategic standpoint on technology in order to condemn it. He sees it exclusively as a system of control and overlooks its role in the lives of those subordinate to it.

The tactical standpoint of those subordinates is far richer. It is the everyday lifeworld of a modern society in which devices form a nearly total environment. In this environment, the individuals identify and pursue meanings. Power is only tangentially at stake in most interactions, and when it becomes an issue, resistance is temporary and limited in scope by the position of the individuals in the system. Yet insofar as masses of individuals are enrolled into technical systems, resistances will inevitably arise and can weigh on the future design and configuration of the systems and their products.

Consider the example of air pollution. So long as those responsible for it could escape the health consequences of their actions to green suburbs, leaving poor urban dwellers to breath filthy air, there was little support for technical solutions to the problem. Pollution controls were seen as costly and unproductive by those with the power to implement them. Eventually, a democratic political process sparked by the spread of the problem and protests by the victims and their advocates legitimated the externalized interests of the victims. Only then was it possible to assemble a social subject including both rich and poor able to make the necessary reforms. This subject finally forced a redesign of the automobile and other sources of pollution, taking human health into account. This is an example of the politics of holistic design that will lead ultimately to a more holistic technological system.

An adequate understanding of the substance of our common life cannot ignore technology. How we configure and design cities, transportation systems, communication media, agriculture, and industrial production is a political matter. And we are making more and more choices about health and knowledge in designing the technologies on which medicine and education increasingly rely. Furthermore, the kinds of things it seems plausible to propose as advances or alternatives are to a great extent conditioned by the failures of the existing technologies and the possibilities they suggest. The once-controversial claim that technology is political now seems obvious.

Culture

Philosophy of technology demystifies the claims to rational necessity and universality of technical decisions. In the 1980s, the constructivist turn in technology studies offered a methodologically fruitful approach to demonstrating this in a wide range of concrete cases. Constructivists show that many possible configurations of resources can yield a working device capable of efficiently fulfilling its function. The different interests of the various actors involved in design are reflected in subtle differences in function and preferences for one or another design of what is nominally the same device. Social choices intervene in the selection of the problem definition as well as its solution. Technology is socially relative and the outcome of technical choices is a world that supports the way of life of one or another influential social group. On these terms the technocratic tendencies of modern societies could be interpreted as an effect of limiting the groups able to intervene in design to technical experts and the corporate and political elites they serve.

Constructivism presupposes that there are many different solutions to technical problems. Some sort of meta-ranking is therefore necessary to choose between them. In determinist and instrumentalist accounts, efficiency serves as the unique principle of meta-ranking. But contemporary technology studies contests that view and proposes that many factors besides efficiency play a role in design choice. Efficiency is not decisive in explaining the success or failure of alternative designs since several viable options usually compete at the inception of a line of development. Technology is "underdetermined" by the criterion of efficiency and responsive to the various particular interests that select among these options.

In my formulation of this thesis, I argue that the intervention of interests does not necessarily reduce efficiency, but biases its achievement according to a broader social program. I have introduced the concept of "technical code" to articulate this relationship between social and technical requirements. A technical code is the realization of an interest in a technically coherent solution to a problem.

Where such codes are reinforced by individuals' perceived self-interest and law, their political import usually passes unnoticed. This is what it means to call a certain way of life culturally secured and a corresponding power hegemonic. Just as political philosophy problematizes cultural formations that have rooted themselves in law, so philosophy of technology problematizes formations that have successfully rooted themselves in technical codes.

This account helps to understand the nature of real world ethical controversies involving technology. Often these turn on the supposed opposition of current standards of technical efficiency and values. I have tried to show that this opposition is factitious, that often current technical methods or standards were once discursively formulated as values and at some time in the past translated into the technical codes we take for granted today. This point is quite important for answering the usual so-called practical objections to ethical arguments for social reform. It is also relevant to the corresponding appeals to ideal values in opposition to the materialism of our society such as we find in Borgmann's chapter here.

The larger implication of this approach has to do with the ethical limits of the technical codes elaborated under the rule of operational autonomy. The very same process in which capitalists and technocrats were freed to make technical decisions without regard for the needs of workers and communities generated a wealth of new "values," ethical demands forced to seek voice discursively. Most fundamentally, democratization of technology is about finding new ways of privileging these excluded values and realizing them in the new technical arrangements.

Instrumentalization Theory

Essentialism, Constructivism, and Instrumentalization Theory

David Stump and Simon Cooper propose opposite improvements on my theory, while Trish Glazebrook sees a parallel between my approach and eco-feminism. I will treat these three chapters together as they have to do with the dilemma of essentialism to constructivism. Much philosophy of technology offers very abstract and unhistorical accounts of the essence of technology. These accounts appear painfully thin compared to the rich complexity revealed in social studies of technology. Yet surely technology has some distinguishing features and these must have normative implications of some sort. As Marcuse argued in *One-Dimensional Man*, the choice of a technical rather than a political or moral solution to a social problem is politically and morally significant. The dilemma is sharply etched in political terms. Most essentialist philosophy of technology is critical of modernity, even antimodern, while most empirical research on technologies ignores the larger issue of modernity and thus appears uncritical, even conformist, to philosophers of technology.[12]

I find it difficult to explain my solution to this dilemma, as it crosses lines we are used to standing behind. These lines cleanly separate the substantivist critique of technology as we find it in Heidegger from the constructivism of many contemporary historians and sociologists. These two approaches are usually seen as totally opposed. Nevertheless, there is something obviously right in both. I have therefore attempted to combine their insights in a common framework, which I call "instrumentalization theory."

Instrumentalization theory holds that technology must be analyzed at two levels, the level of our original functional relation to reality and the level of design and implementation. At the first level, we seek and find affordances that can be mobilized in devices and systems by decontextualizing the objects of experience and reducing them to their useful properties. At the second level, we introduce designs that can be integrated with other already existing devices and systems and with various social constraints such as ethical and aesthetic principles. The primary level simplifies objects for incorporation into a device while the secondary level integrates the simplified objects to a natural and social environment.

These two levels are analytically distinguished. No matter how abstract the affordances identified at the primary level, they are likely to carry social content from the secondary level in the elementary concept of use. Similarly, secondary instrumentalizations such as design specifications presuppose the identification of the affordances to be assembled and concretized. This is an important point. Cutting down a tree to make lumber and building a house with it are *not* the primary and secondary instrumentalizations respectively. Cutting down a tree "decontextualizes" it, but in line with various technical, legal, and aesthetic considerations determining what kinds of trees can become lumber and are salable as such. The act of cutting down the tree is thus not simply "primary" but involves both levels, as one would expect of an analytic distinction.

Analysis at the first level is inspired by categories introduced by Heidegger and other substantivist critics of technology. However, because I do not ontologize those categories, nor treat them as a full account of the essence of technology, I believe I am able to avoid many of the problems associated with substantivism, particularly its antimodernism. Analysis at the second level is inspired by empirical study of technology in the constructivist vein. I focus especially on the way actors perceive the meanings of the devices and systems they design and use. But again, I am selective in drawing on this tradition. I do not accept its exaggerated and largely rhetorical empiricism and its rejection of the categories of traditional social theory. Instead, I attempt to integrate its methodological insights to a more broadly conceived theory of modernity.

As we will see in a later section, Thompson and Borgmann do not approach the instrumentalization theory in this light. They assume that the primary instrumentalization concerns efficiency, and in some sense this is true, but it would be more accurate to say that it identifies and secures affordances. Affordances are not yet a device and so cannot be efficient or inefficient. The secondary level is not an extraneous imposition of "values" on these affordances but their systematic incorporation into an object that makes sense in both technical and social terms. No technology can exist that does not meet this condition.

Stump thinks I would do better to abandon substantivist critique altogether. He considers my cautious attempt to reconcile some aspects of substantivism with a more empirically oriented approach still insufficiently empiricist. The historical notion of technological momentum captures the valid intuition of thinkers like Heidegger and Ellul better than their essentialist speculations on the nature of technology as such. Cooper worries that I do not place sufficient emphasis on substantivist critique. He fears that I am unable to cope with innovations, such as biotechnology, that alter the very nature of our humanity. Were I to follow Stump's suggestion, I would be even more completely disarmed in the face of the Cooper's critique. Were I to follow Cooper, Stump would have even more worries about my penchant for metaphysics. I would like to consider the arguments of these two chapters in more detail before turning to Glazebrook's illuminating comments.

Momentum

Stump begins by arguing that strict social constructivism is no more useful for a social theory of technology than is substantivist essentialism. The social constructivism he attributes to Bloor holds that technological development is determined exclusively by society without technical considerations, properly speaking, playing any role at all. This is so improbable that I have difficulty accepting that anyone believes it, but the world of science and technology studies is a rich tapestry.

In any case, I agree with Stump that "[t]he concrete setting of technological development always includes social or political, as well as technical or immanent concerns, and all of these need to be included in an account of technological development."[13] Our disagreement on constructivism is largely semantic. Who is a constructivist? I would have thought that Bijker, Pinch, Hughes, and others whose work is published in the two well-known volumes from MIT Press on the social construction or shaping of technology would be counted as constructivists in a loose sense.[14] In this sense, I think the term is useful enough to want to hold on to it despite its vagueness.

Next, Stump moves on to his objections to my lingering fidelity to my substantivist origins. It is true that my instrumentalization theory stands in the place of substantivist theories of the essence of technology and continues to use concepts familiar from that tradition. Stump complains that substantivism defines an essence of technology that is separate from society, teleological, deterministic, and unified. He admits that substantivist theory does have some useful insights into the autonomy of technological systems, but the metaphysical apparatus introduced to explain it is unnecessary. Historians such as Thomas Hughes have developed similar insights on an empirical basis by studying the development of particular technological systems. Hughes's concept of technological momentum is a more precise and empirically grounded way of addressing substantivism's concerns.

This conclusion has far-reaching implications. Modernity theories emphasize the differentiation of society and the autonomy of technology while technology studies emphasizes the social and political conditioning of technological choices. The concept of momentum reconciles these apparently opposite perspectives by showing how socially and politically conditioned technological institutions gradually gain independence of their environment and become relatively autonomous in the course of development. By sticking close to the empirical facts, this sort of historical analysis grounds generalizations in experience rather than in a priori speculation.

Stump argues that a methodological commitment to empirical grounding should also guide us in evaluating political claims. The appeal to alternatives to existing systems must be based on real possibilities for change, not on the abstract logic of underdetermination. But in many cases real possibilities are so limited by technological momentum as to be all but nonexistent. Preserving the contingency of development in general is not much use in such cases. Stump's political conclusion is thus more pessimistic than mine, not for Heideggerian essentialist reasons but rather on the basis of empirical observation of the robustness of large-scale systems.

Posthumanism

Simon Cooper's argument is the mirror image of Stump's. Stump wants me to abandon essentialism altogether since large-scale technical systems do everything that substantivist theory can plausibly attribute to technology as such. Cooper argues that I am not enough of a substantivist to address the problems raised by contemporary biotechnology.

According to Cooper, biotechnology presents a posthuman challenge. It transforms the meaning of our activities and of our very selves so profoundly that it calls into question the ideals of citizenship and democratic control to which I appeal. Reproductive technologies, for example, do not simply enable infertile couples to have children; more importantly, they alter the meaning of childbearing by freeing it from the "limitations of embodiment." Cooper argues that the cultural and social context of our humanity is thinned out and redefined by new technologies. The strong notion of human potentialities to which I appeal for a normative foundation cannot support the weight I make it bear since what it is to be human is transformed by the very technologies I measure by its standard.

He recognizes the risk that he will fall into essentialism and determinism in pursuing this line of argument and he hopes to avoid such theoretical backsliding. He wants to reinstate the radicalism of Heidegger's questioning of technology not with an essentialist and determinist argument, but through a consideration of the contingent cultural bases of our humanity. These historically evolved criteria of what it is to be human are altered by biotechnology. In addition, the forms of agency traditionally associated with face-to-face human contact are subverted by electronic mediations. Cooper wonders if agency can be realistically attributed to the "specific intellectuals" working in this new electronic setting that favors extreme individualism. The cultural shift toward "a looser, more abstract mode of being-in-the-world" undermines the sources of cooperation and agency so deeply that we must address this shift as the principal issue of a politics of technology.

Mutually Cancelling Criticisms

I do not agree that I have overlooked the issues Cooper raises. I have introduced a historical conception of humanism and applied it to several case studies. The discussions of experimental treatment in *Alternative Modernity* and of on-line education in *Transforming Technology* illustrate how I deal with what might be called the "boundary problem," that is, defending a certain conception of humanity from destructive technical mediation.

The instrumentalization theory provides the basis for addressing the issues Cooper raises. The level of primary instrumentalization demarcates technical action from other modes of action. In specifying the essence of technology it also suggests technology's appropriate limits. Once one knows that technical action involves such operations as decontextualization and reduction, it should be clear that not everything is a proper object of technique. There are many human relationships, for example, that would simply dissolve if managed technically. What would become of a friendship if it were addressed from the standpoint of the affordances it offered? Could a child decontextualized in a "Skinner box" be properly reared?

Secondary instrumentalizations are able to mitigate many of the effects of the primary instrumentalization but not erase them completely. We are not as badly off as the Native American praying to the soul of the buffalo he has killed, but there are definite limits to what can be restored by design work. Technology's appropriate reach can be defined by reference to this schema. Where the benefits of functionalization outweigh the human consequences, we go forward. Where it threatens our humanity, we should learn to pause, reflect, and abstain.

This general point needs to be concretized in terms of Cooper's two technological examples, biotechnology and electronic mediation. But I cannot tell from his comments whether he would block these technologies altogether or attempt to control and design them to conform to our historically evolved standards. Abandoning these technologies completely makes no sense. The slippery slope arguments employed to justify such an extreme reaction have no logical stopping point. If we block reproductive technologies that help infertile couples have babies, then why not block abortion and contraception as well? What about the medicalized delivery of babies? Should we get rid of it too? Restoring the total contingency and real danger of pregnancy is not progress.

So we are left with the alternative of controlling the technologies and designing them in the most humane way possible. We can be guided in this task by an understanding of the limits of technology. The substantivist elements of the instrumentalization theory help to identify those limits. To give a simple example, we could determine on this basis that the choice of such characteristics of the fetus as gender should be forbidden. This type of choice represents an illegitimate reduction of the human individual to a socially preferred "primary quality." A similar argument could be made concerning electronic mediation and its impacts on individuality and agency. We should oppose the reduction of the user of the computer to a mere consumer on an Internet functioning as an electronic mall in favor of designs that offer opportunities for human contact and free assembly.[15]

Now I can show the problem with Stump's rigorous antisubstantialism. While the theory of large-scale technical systems does address one of the principal insights of substantive theory in an empirically grounded fashion, it does not help with the problems Cooper identifies. Some conception of the nature and limits of technology is required to judge its appropriate place in social life. Without such a conception, there are no specifically technical reasons for controls on deployment and design. One might, for example, still object to gender selection on the grounds of women's rights. But one could not claim that technology is inappropriately applied or designed to produce the effects that violate women's rights. This complementary argument is significant given the overwhelming impact of technology in every aspect of contemporary life. As Cooper argues, it is important to be able to address technology in terms of the conditions of our humanity, rather than treating it as a merely neutral instrument in service to separately conceived values.

The Feminist Parallel

Trish Glazebrook's chapter brings out the significance of this distinction. She finds interesting parallels between my theory of technology and contemporary eco-feminism. Just as I call for a cultural hermeneutic of technology, eco-feminism calls for a cultural approach to gender. Second wave feminism could be thought of as "a subversive rationalization of gender." Just as feminism strives to liberate women from subjection to their own bodies, so critical theory of technology "argues for the possibility of an alternative modernity in which human beings participate in technology as rational agents, that is, self-determining subjects, rather than simply being subjected to technical codes and practices controlled by an elite group."[16]

Glazebrook argues that the nature/nurture distinction, like the essentialism/constructivism distinction, must be overcome. Technologies are not reducible to causal mechanisms any more than women are to biological ones. Efficiency, like sex, is a very partial description of devices and bodies. Instead of opposing biology to culture, contemporary feminists view woman's bodies as a political site, a terrain of struggle. Biology is not irrelevant to gender identity—gender is not purely contingent—but neither does biology explain it. Similarly, causal and symbolic aspects of technology coexist and interpenetrate.

Critical theory of technology overcomes the opposition between nature and culture in the technological domain. The instrumentalization theory combines essentialist insights into functionalization with a constructivist understanding of meaning and context. Many feminists would agree with critical theory of technology that a historical approach can overcome such oppositions. The natural characteristics of women's bodies are seamlessly bound up with social practices and meanings in a way that tends to deny history. But careful analysis brings history back in as the scene on which practices and meanings are generated. The form of the body itself is altered in response as different social and economic relations, and different aesthetic ideals remold it within its biological limits. Simply put, women continue to be the sex that has children, but pregnancy and childbirth may be different practices with different meanings at different times. Something similar happens to technology, although it has a much larger space for variation. Historical development supplies new practices and meanings which lead to changes in technology, which, of course, retains a technical coherence and rationality in whatever form it adopts.

These reflections help to contextualize the difference between Stump and Cooper. They each challenge me to explain extreme consequences of technology. Stump suggests a different approach to the critique of technocracy. Cooper points to the human consequences of technization. It is true, as Stump argues, that historically contingent developments such as large-scale technical systems must be understood in their empirical complexity rather than deduced from an essence of technology. But it is also true, as Cooper argues, that technical systems set in motion a specific type of change when they extend into regions that were formerly sheltered from technical control.

We need a method that recognizes both sides of the argument. This is what I hope to accomplish with the instrumentalization theory. The history of large-scale technical systems falls under the secondary instrumentalization I call "systematization." The empirical approach to these systems fills out the instrumentalization theory rather than contradicting it as Stump assumes. Similarly, the primary instrumentalizations of decontextualization and reduction undergird the critique of the dehumanizing applications of technology that concern Cooper.

Glazebrook points out that this approach resembles Heidegger's historical notion of essence, which "is about location and situatedness." History "disrupts technological essentialism and determinism" and makes it impossible to define technology in terms of causality alone.[17] While I can agree with this, I would argue that Heidegger does not have a concrete understanding of the social and symbolic aspects of design. Pursuing this line would further destabilize the naturalization of technology by revealing its political significance. Feminism has of course done precisely this for women's bodies. I develop this criticism of Heidegger in the next section.

Heidegger

Heidegger's Promise and Limitations

Glazebrook and Thomson believe that Heidegger's theory of technology can guide us today. I am not surprised that Thomson, who studied with Hubert Dreyfus, should come to the defense of Heidegger. Dreyfus himself has written several interesting articles in which he attempts a similar salvage operation.

Like Dreyfus, Thomson refers us to a passage in Heidegger's essay, "Building Dwelling Thinking," where the modern highway bridge functions as a "thing" in Heidegger's eminent sense of the term.[18] It is true that in this passage Heidegger discusses modern technology without negativism or nostalgia and suggests an innovative approach to understanding it. Combining this unique example with his many obscure and ambiguous statements on technology in general, one can construct connections between Heidegger and Woodstock, as does Dreyfus, or, more plausibly, Heidegger and environmentalism, as Glazebrook suggests here. But how plausible are these interpretations, really?

The problem is Heidegger's seeming indifference toward the design of technology. Thomson is right that the adjective "unhistorical" does not quite apply to Heidegger's theory. What I called "unhistorical" about his account of modern technology is not that it lacks an origin, but that it lacks a future. I can find no indication in his thought that the *things* we normally refer to as "modern technology" can be significantly improved. Even if the "mode of revealing" were to shift away from the technological enframing, we would still use the same devices.

We live in a society in turmoil around technical issues in communications, computers, medicine, the environment. How are we to intervene and for what? I have argued for an appreciation of the role of the technical lifeworld in which we live with devices, not merely controlling them but also finding meaning through them. This approach draws on Heidegger's philosophy of art rather than his philosophy of technology. Heidegger describes works of art as setting up "worlds," realms of meaning that inform the life of a community. The Parthenon does more than shelter worshippers; it lays out the space of the city with itself as center. I believe technologies hold a similar place in modern life. The opposition in Heidegger between the worlding power of art and the deworlding effects of technology must be overcome to devise a critical theory of technology.

The Historical Concept of Essence

The philosophical core of our disagreement concerns the relation between the ontic and the ontological in the understanding of technology. Thomson emphasizes that Heidegger's essence of technology, "enframing" (*Gestell*), refers to the ontological rather than the ontic level. What Heidegger calls "technology" we would more likely call an attitude in which everything appears as a resource. Heidegger's claim that we live in a technological age would then be roughly equivalent to the notion that modern culture comprehends everything as a potential object of technical action.

The ontic, by contrast, is the level of empirical objects, of actual machines and the nature they transform, of our own needs and activities, hence also of ordinary political strife and struggle. The "ontological difference" appears to insulate the one from the other. It is true, as Thomson points out, that Heidegger made an exception for the Nazi revolution, but I find no theoretical basis for this exception in his thought. Certainly, he never claimed that ontic struggles over the design of devices could change the ontological dispensation within which the world appears as technological.

Thomson follows Heidegger in claiming that the essence of technology is not a genus under which modern technologies would fall as particular instances but an ontological happening of some sort. This seems rather confusing at first glance: What sense does it make to call something an essence if it is not a genus? Yet the conclusion that both Glazebrook and Thomson draw, namely, that Heideggerian essences are historical and dynamic, is one I would like to defend too on somewhat different grounds.

I believe there is a way to show that enframing is at least not *primarily* a genus in the usual sense. Consider the parallel case of culture or language. Culturally encoded behavior or speech are not particulars in the same way in which, for example, red paint is an instance of the genus red or a coffee cup an instance of the genus cup. Culture and language are enacted, and the enactments reproduce them concretely rather than simply instantiating them. Culture and language are thus what Hegel called "concrete universals"—they exist in their instances—in contrast with abstract universals that generalize from particulars. Heidegger indicates that this is the sort of distinction he wants to make when he says, "If we speak of the 'essence of a house' and the 'essence of a state,' we do not mean a generic type; rather we mean the ways in which the house and state hold sway, administer themselves, develop and decay—the way in which they 'essence' [*Wesen*]."[19]

Thomson argues that Heidegger's historical concept of essence can do the work of a critical theory of technology. It does suggest a connection between levels such as that which joins the flow of spoken language or cultural practices to the gradual changes in the structures they enact. In that case the theory of enframing cannot be cleanly separated from an analysis of particular technologies. But then, Heidegger's many reactionary attacks on modern devices become all the more significant. I would like to conclude with an example I find particularly revealing.

In 1962 Heidegger gave a speech in which he explained the difference between language as saying, as revealing the world by showing and pointing, and language as mere sign, transmitting a message, a fragment of already constituted information. The perfection of speech is poetry, which opens language to being. The perfection of the sign is the unambiguous position of a switch—on or off—as in Morse code or the memory of a computer. Heidegger writes that with the introduction of the computer, "[t]he type and style of language is determined according to the technical possibilities of the formal production of signs, a production which consists in executing a continuous sequence of yes-no decisions with the greatest possible speed. . . . The mode of language is determined by technique."[20]

This makes fun reading for philosophers, but it is embarrassingly wide of the mark. What has actually happened to language in a world more and more dominated by computers? Has it been reified into a technical discourse purified of human significance? On the contrary, the Internet now carries a veritable tidal wave of "saying," of language used for expression as always in the past. The simple fact of the case is that these "posthumanist" reflections on the computer were wrong. They not only failed to foresee the transformation of the computer into a communication medium, but they precluded that possibility for essential reasons.

Ah, but was the error ontic or ontological? In considering this question it becomes clear why the wall between the two realms breaks down. Underlying the ontic analysis of the computer there is an ontological presupposition according to which technology introduces a peculiarly impersonal form of domination into human affairs. This presupposition is then played out at the ontic level in the seeming enactment of impersonality and control in the unambiguous positioning of the digital switch. The resulting "aggression of technical language against the proper character of language is at the same time a threat against the proper essence of man."[21] Now we are returned to the ontological level. The ontological appears in the ontic; the ontic strikes back at the ontological. The two are linked in Heidegger's discourse, not separate, as many of his interpreters claim.

Because of this subterranean linkage, ontological presuppositions intrude unacceptably on the ontic level. That is the source of the erroneous evaluation of the computer. The chain of equivalences, which runs from the impersonality and domination of technology as such down to particular devices such as computers, gets in the way of concrete analysis. A serious encounter with particular technologies shows that they have many dimensions that can be actualized under different social and historical circumstances. Technology has never had a single meaning, such as enframing, which summed up all its potentials. The ability of the computer to mediate natural language is not a startling reversal of ontological trends, but merely an expression of the complexity and flexibility of technology that is revealed as it is appropriated by a wider range of actors than the scientists and engineers who employed it in 1962.

What conclusion do I draw from these reflections? I do think Heidegger's philosophy of technology is interesting and suggestive. While I can borrow the concept of world from his philosophy of art and find resources for understanding the primary instrumentalization in his philosophy of technology, we need to look to other sources for an understanding of design and technological activism.

The Politics of Technology

Technology and Rights

I turn now to questions surrounding the democratization of technology. I begin by addressing Doppelt's chapter, which typifies the main dissatisfactions of political philosophers with philosophy of technology, especially my own. Doppelt argues that I lack a "clear and plausible standard of what counts as the democratization of technology," that I fail to offer a "substantive conception of democratic ideals," and that I do not adequately take into account arguments in favor of the undemocratic management of technology based on property rights. In sum, while my demystification of technocratic ideology is successful, I cannot get from there to an adequate account of the democratization of technology. The main reason for this failure is my lack of an explicit equality-based argument justifying technical democratization and differentiating it from other non- or undemocratic forms of technical change.

The background to Doppelt's argument is the distinction political philosophers make between merely particular interests and rights that are rationally grounded and can claim universality. Interests and rights are not mutually exclusive concepts. Usually a right protects or furthers an interest, often in conflict with other interests that cannot be universalized. For example, the interest of the disabled in free movement around the city is interpreted as a right in current legislation, and this right overrides the interest of taxpayers who are now obliged to pay for the retrofitting of sidewalks and public buildings for disabled access.

When this distinction between interests and rights is brought to bear on the arguments of philosophers of technology such as myself, we appear to fail an important test of rigor. Why, after all, should we prefer a world in which workers have more rights and businessmen less, in which certain risks such as those associated with nuclear power are taken more seriously than the benefits, and so on? All these preferences involve clashes of interests that, if they cannot be reconciled by compromise, ought to be adjudicated in terms of rights. But philosophers of technology rarely invoke the discourse of rights or argue for the rights corresponding to their preferences.

Doppelt notes that my explicit arguments for democratizing technology are based on the value of agency, but agency by itself, without some substantive goal sanctioned by the ideal of equality, does not constitute a democratic value.

In fact, the successful assertion of agency by some groups may disempower and oppress others and violate democratic norms. Since I do not present a "standard of justification" to sort out these cases my theory is incomplete.

Doppelt's critique is interesting and suggestive, but it relies heavily on the tradition of political philosophy, which has its own problems. Politics is a form of being that only exists by the common consent of human beings and that joins them in a community. The political lies between us; we bring it into being by abstaining from violence in the resolution of our civic differences, and it in turn constitutes us as citizens. This unusual mediation requires explanation. Unfortunately, the tradition of political theory has been fixated for centuries now on a mythic account of the political, the social contract. But the political is not a self-sufficient form of life; politics is founded in many other forms of "between-being" uniting men and women in a society. Without an understanding of these social relations, political relations remain obscure.

Political philosophy has struggled in recent years to face this issue. It would still like to abstract the political bond from social bonds so as to focus the significance of justice in human affairs more sharply than is possible in the muddle of everyday social life. Instead of returning politics to its social roots, political philosophy has attempted to incorporate aspects of social life into a framework still implicitly determined by the social contract. This has led to a flowering of theories of poverty, gender, and race. I argue that the project of understanding the links between the social forces shaping our society and its ideals of freedom, justice, and equality cannot be carried through without also taking account of technology.

This view is not generally accepted in political philosophy. Consider the difference between the treatment of income inequality and the treatment of differential access to control over technical arrangements. Most political philosophers agree that beyond a certain point material inequalities undermine democratic rights. Hence the welfare state in some form has been integrated to the conception of justice, notably by Rawls.

However, technical arrangements are also often disempowering and oppressive. Workers in many dead-end jobs are subjected to injuries to physical and mental health and to their dignity as human beings as surely as those whose income falls below the poverty line. Had these workers more control over their work, they might use that control to make it safer and more fulfilling. But the typical organization of work in a modern industrial economy leaves them little if any initiative. Why would one worry about the rights of the poor in income and not the rights of the poor in control over the material circumstances of their work? A political philosophy conscious of the problem of excessive income inequality ought to address these issues as well. Similar issues arise around the control of the media of communication.

Surely differences in control have political implications as least as large as differences in income. Yet political philosophy has paid little attention to the obvious injustices that arise when a few can purchase the space of public debate, reducing the many to manipulated objects.

Doppelt is in fact one of the rare political philosophers to have squarely faced the problems of working life. He argues that human dignity in the workplace is a matter of human rights and should be protected by a just society.[22] However, his argument is incomplete so long as it is not grounded on a philosophy of technology capable of rendering a greater measure of economic justice practically plausible. After all, as Borgmann points out in his chapter, if respect for workers will impoverish our society, even workers are unlikely to insist on their rights.

The Humanistic Tradition

Political philosophers disagree over which aspect of democratic right to emphasize, citizen agency or equal individual liberties, autonomy in public life or in private life. These two emphases correspond to two great and influential traditions of democratic thought, which Isaiah Berlin summed up rather misleadingly with his concepts of "positive" and "negative" liberty. Although this is a cogent distinction, it misses the main thrust of the critical philosophy of technology that has emerged over the last twenty years.[23] This philosophy invokes a third conception of liberty found in what I will call the humanistic tradition. This tradition holds that a good society should enable human beings to develop their capacities to the fullest.

When Albert Borgmann or Langdon Winner criticizes current technical arrangements, it is primarily in terms of the limitations those arrangements place on human development. Technology in their view provides the material framework of modern life. That framework is no neutral background against which individuals pursue their conception of the good life, but instead informs that conception from beginning to end. The most important question to ask is what understanding of human life is embodied in the prevailing technical arrangements.

It may be objected that the very notion of "capacities" is normative and so must itself be grounded. Although the objection appears daunting, I do not think political philosophy has it much easier facing similar demands for justification of its values of agency and equality. After all, how do we distinguish "true" agency from manipulation? And what makes some uses of our private freedom politically worthy and others condemnable? In all such cases, criteria are needed. Political philosophers often try to derive these criteria from Kantian or Utilitarian theories of moral obligation, or from ideal constructions of rational discourse. I do not find any of these arguments particularly compelling.

The humanist tradition is supported by a different kind of argument, which has a precedent in Hegel. On this account, grounds are derived from the potentialities revealed in a historical tradition to which we belong. This Hegelian humanism seeks evidence in history that our destiny as human beings is a progressive unfolding of capacities for free self-expression, the invention of the human. Because we belong to the tradition shaped by struggles to improve the human estate, wherever we see similar struggles for a fuller realization of freedom, equality, moral responsibility, individuality, and creativity we interpret them as contributing to the fuller realization of human capacities. But note that while the notion of realizing human capacities refers us to a totality of the human, that totality is not given in advance in a speculative ideal but must emerge from the real process of struggle, piece by piece.

I stated earlier that politics involves a special kind of social bond and is embedded in a wider framework of social bonds. Increasingly, these bonds are technically mediated in our society. This is why a concept of politics that abstracts systematically from technology will prove more and more irrelevant to many of the most compelling problems we face. Philosophy of technology may well have something to learn from political philosophy, but political philosophy must make a similar opening to philosophy of technology and broaden its reach to encompass the realities of our time.

Democratic Interventions

Criteria for Reform

In *Questioning Technology* I develop a theory of democratic interventions and outline three types of technical micropolitics, controversies, public participation in design, and reinvention of technologies by users. I call these "democratic rationalizations" because they involve public participation in the realization of technically coherent designs. The three types of intervention I explore do not constitute an exhaustive list but rather show the possibility of a radical politics of technology. Albert Borgmann's chapter evaluates the plausibility of my argument. Thompson, Light, and Woodhouse all suggest useful additions and amplifications. Light emphasizes the role of democratic participation in environmental reform, Thompson the importance of decommodifying strategies, and Woodhouse the role of government in establishing institutions favorable to publicly responsible technological development.

Borgmann lays out "[t]he criteria for real reform . . . political realism, cultural depth, structural comprehensiveness, and substantive content."[24] Since philosophy of technology has a Left coloration, it attempts to find technological movements that resemble such influential models as the civil rights movement or feminist movement, the very movements I identify with the humanistic tradition. But these movements do not address technology and nothing quite like them has emerged to fulfill the criteria of real reform. Technological reform may require different approaches that do not resonate with Left assumptions. Borgmann does not say what these are here but I assume from his other writings that they have to do with civic life in small face-to-face communities.[25]

The four criteria appear difficult to meet. Realism usually excludes critical depth. Proceduralism is a temptation of philosophers faced with substantive issues they do not know how to address. It is difficult to separate the criteria and so to define them independently. According to Borgmann, my critical theory of technology "breaches this vicious circle" by identifying the political indeterminacy of technology and finding a place for the actions of ordinary people in the technical domain. Feasibility is demonstrated by the examples of change I document, sweep and scope are addressed in terms of an empirical orientation toward the complexity of technological cases, and depth is made possible by the attention to actual nuts and bolts workings of technology.

I appreciate Borgmann's analysis of my accomplishment in meeting his four criteria but he appears to take away with one hand what he grants with the other. He suspects that technological reform may not be popular enough to have the impact I hope for. He discusses this problem in terms of an interpretation of my instrumentalization theory with which I have some disagreements. I note that Thompson shares a similar interpretation, which suggests that I have failed to explain my theory adequately. I will try briefly here to clarify certain points.

The instrumentalization theory as Borgmann and Thompson understand it concerns the relation between the pursuit of efficiency and social values, the former identified with the primary instrumentalization and the latter with the secondary instrumentalization. An emphasis on the primary instrumentalization thus appears linked to higher levels of consumption and prosperity. In Borgmann's interpretation, an emphasis on the secondary instrumentalization leads instead to a life elevated by high aesthetic and moral standards. Which values are most influential? Borgmann is not happy with the obvious answer to this question. "One obstacle, then, in the path of democratic technology may be people's preference for affluence over autonomy."[26]

Now, it is true that many of my examples of secondary instrumentalizations soften the hard edges of technological interventions into social life. But secondary instrumentalizations are not necessarily based on higher ideals. They often respond to quite banal technical or social requirements. When I argue that capitalism inhibits certain secondary instrumentalizations protecting human beings and nature from exploitation, I should have made it clear that the primary instrumentalization alone cannot guide capitalist technical design. Some sort of secondary instrumentalizations are necessary to integrate technology to any society, a bad one as well as a good one.

I am not ready to concede the alternative of autonomy and prosperity with which Borgmann concludes his chapter. I have devoted a good deal of effort to refuting this alternative, which I call the "trade-off" theory of technological reform. In opposition to this theory, I argue that the design process often combines goals such as environmentalism and industrial production, medical experimentation and patient rights, informational and communicative applications of computers, and does not necessarily require a choice between them. This is the argument of my theory of technological concretization, discussed in the conclusions of *Questioning Technology* and *Transforming Technology*. In sum, I hold out the hope that technological development can reconcile us rather than divide us, on condition that democratic interventions succeed in orienting innovation toward new types of designs that incorporate secondary instrumentalizations blocked by the capitalist technical code.

Commodification

I turn now to Paul Thompson's chapter. Thompson suggests that my theory could be improved by incorporating the concept of commodification. I agree with his general point and see the interest of the connections he builds between my theory, Borgmann's, and, surprisingly, Latour's.

Commodification in Thompson's account involves four processes that transform goods into exchangeable objects. Goods must be made "alienable," separable from each other and their owners, "excludable," easily monopolized by an owner, "rival," in the sense that they are good for only one use, and, finally, they must be standardized. The extent to which a good fulfills these four criteria determines the degree of its commodification.

Goods are commodified in two ways, either "structurally," by legitimating the four criteria, or "technologically," by appropriate designs. For example, music was subject to structural commodification (concerts, sheet music) before its technological commodification in sound recording, which made possible its alienation from the performer.

Thompson believes the theory of technological commodification offers "a more detailed account of secondary instrumentalization in some of its most influential and enduring forms."[27] He interprets secondary instrumentalizations as anticapitalist strategies of technological development. The AIDS case I analyze in *Alternative Modernity* can be explained on these terms as a decommodification of care. The medical community assumed that care would be delivered in the commodified form of drugs but patients also demanded personal care. The Minitel case

I present in the same book is another example of decommodification, which continues with the Internet. In both systems human communication proceeds uninhibited and outside any exchange relation, frustrating the expectations of information society bureaucrats and dot-com businessmen.

Since Thompson interprets my distinction of primary and secondary instrumentalization in terms of capitalist versus anti-capitalist programs, he finds no place in it for "agency" in Latour's sense. Thompson claims that my "philosophy of technology is not equipped with ideas that allow us to understand technology itself as active in shaping human intentions and human relationships, and this omission limits the rapprochement with science studies."[28] Technological commodification can serve as an analytic category for interpreting the agency of things since it has many unintended and uncontrollable consequences. Technologies thus appear as "political actors" in Thompson's revision of my theory.

I confess I am puzzled by this criticism given the extensive discussions of environmentalism in several places in my work, notably in *Questioning Technology*, where Barry Commoner's account of the unintended consequences of post–World War II industrialization is described approvingly in some detail.[29] Where I do disagree is over Latour's symmetrical description of the "agency" of things and persons. Thompson's theory of technological commodification does not seem compatible with this strange notion in any case. Surely human agents must initiate the commodifying and decommodifying, in other words, set in motion processes that involve the attribution of meanings to things and which may have unintended consequences. Interpreted in this way, I can accept fully Thompson's amplification of my approach.

Participation

There is an interesting connection between Thompson's approach and Light's. Light worries that I do not discuss the process of environmental reform beyond noting that issues such as pollution may motivate demands for technological change. The place of democracy in environmental reform is surely more complex than this. Light distinguishes between environmental reforms that achieve substantive goals within an undemocratic social framework and those that are designed to favor continuing democratic participation. In Thompson's terms, the latter would be less technologically commodified.

Might democratic environmental movements lead to the design of technologies that themselves depend on continuing public involvement? Or is there no connection between the democratic character of the movement and its technological outcome? I do not directly address these questions but tend to assume in this as in other cases that, overall, democratic movements for technological change will lead to more democratic designs. I will call this the "continuity thesis," according to which democracy is not only a means of change but a result of change, procedure, and substance in one. Only in the discussion of labor-management relations in *Transforming Technology* have I offered anything like an analysis of the interconnection of democratic movements for technological change and their outcomes. In that case it seems plausible to argue that workers who actively participate in technological choices will be less likely to favor disempowering and deskilling alternatives than will managers who benefit from such designs.

Environmentalism suggests another type of argument for the continuity thesis. Light claims that "as a human technological practice that involves our cultural connection to nature, we need to think about what values are produced in that practice and how those values can best be made use of in the broader, long-term project of creating a sustainable society."[30] Public participation in environmental processes can create a constituency for environmentalism in a way merely changing technology cannot. Environmental restoration is a domain in which Light's approach is particularly meaningful. Engaging volunteers in the actual work of restoration enables people to recognize themselves as "stewards of the environment."[31] This is perhaps a partial response to Borgmann's concerns.

Institutional Reform

I turn now to Woodhouse's comments, which are directed at institutional reforms in support of a more democratic technological system. Like Stump, Woodhouse argues that my examples of the malleability of technology, AIDS experimentation and the Minitel, are too marginal to carry the theoretical burden of refuting determinism. I do not rely on these two examples alone, but in *Transforming Technology* develop a third example, on-line education, in considerable detail. I also discuss environmentalism in *Questioning Technology*. The fact that Woodhouse himself can come up with persuasive examples from the nuclear power and chemical industries indicates that I am on the right track.

Woodhouse's constructive discussion of institutional reform is interesting, although some of his proposals are more plausible than others. He believes that executives are far more influential than other actors in technological decision making and argues that they must therefore be given incentives to make better choices. In a general way I do agree but I see the source of change differently. I discuss the problem of executive power under the heading of "operational autonomy," the discretionary power of managers to influence outcomes in ways that reinforce their own power in the future. In the technological realm this has meant deskilling workers to reduce their autonomy and cheapen their labor. I have proposed industrial democracy as a possible trajectory of development weakening the operational autonomy of management. The difference between us is not that I implausibly suggest that workers in a single company could legislate for a whole country, but rather that Woodhouse thinks government policy rather than labor struggle will be the source of reforms limiting managers' power.

Unfortunately, in the light of recent revelations about corporate governance and the feebleness of reform efforts, Woodhouse's suggestion appears almost as utopian as mine. More radical reforms such as I propose may be unlikely anytime soon, but were they to occur, they would have the virtues Light attributes to participatory environmental reforms and would therefore be sustainable over the long run.

I am more convinced by Woodhouse's idea of a system of public wholesalers. Quasi-public organizations buying in quantity could impose a wider range of goals on manufacturers than private wholesalers. Such a system has been implemented for medicine in many countries. The government as insurer and buyer negotiates on behalf of consumers of medical services and products. Where the system is successful, the government imposes values related to health and quality of life that may be ignored or downplayed by private insurers. Extending such a system widely, perhaps through competing public wholesalers to ensure quality, could have a major impact on technological decision making. One obvious place to start would be energy, for which there is already some precedent. For example, during the California energy crisis, the state created an energy buyer to deal with private producers. The "Independent System Operator" was charged only with keeping the lights on, but it could have been granted a broader environmental mandate.

Dewey

The Rhetoric of Science and Technology

In this concluding section, I will discuss Larry Hickman's remarks on my work here and in his recent book *Philosophical Tools for Technical Culture*. Hickman claims that I have unwittingly reproduced Dewey's philosophy of technology. I am supposedly a pragmatist without knowing it! Now there is no dishonor in being a pragmatist, and Hickman does find some significant similarities between my position and Dewey's. Like Dewey, I reject technological essentialism and determinism for a social theory of technology. And like Dewey I am concerned to preserve and enhance democracy in what Dewey called "the machine age."

But I would be surprised to learn that Hickman was entirely correct. After all, my generational experience and sources are quite different from Dewey's.

Dewey was born in the nineteenth century and elaborated his views "between Hegel and Darwin." I arrived at my approach starting out from Heidegger and the Frankfurt School, that is, from doctrines keenly sensitive to the failure and indeed the threat of modern technology, the famous "dialectic of Enlightenment." The constructivist revision of this theory in my work does bring me closer to Dewey, but Hickman's claims are quite ambitious. He argues that the critical contribution of Heidegger and the Frankfurt School is either misguided or anticipated by Dewey, whose "general pattern of inquiry . . . is capable of absorbing the agenda of the critical theorists. . . ."[32]

If this is so I am indeed a Deweyite. But the problem for me in accepting Dewey as an inspiration has always been his rhetorical celebration of science, technology, and American liberalism. Consider Dewey's argument in *Individualism Old and New*. There he refutes various critics of science and technology, including the New England "genteel tradition" and European aesthetes who defend values more appropriate to an earlier stage of civilization. They have missed the point, which Dewey succinctly states: "In the end, technique can only signify emancipation of individuality, and emancipation of individuality on a broader scale than has obtained in the past." And he states, "It is a curious state of mind which finds pleasure in setting forth the 'limits of science.' For the intrinsic limit of knowledge is simply ignorance."[33] The idea that science might border at its limit on other forms of knowledge does not seem to occur to him here.

It is easy to sympathize with Dewey's aggravation at the indifference, even hostility, of humanistic intellectuals to the great transformations taking place around them. But surely the comments I have quoted are unfair. Technical progress may be *ultimately* liberating but in the present its consequences are ambiguous and this should be a focus of concern, not defensiveness. And the idea that scientific knowledge has limits is not to be confused with a preference for ignorance. Other ways of knowing must fight for their rights in a world dominated by scientistic prejudices Dewey elsewhere deplores.

The consequences of Dewey's excessive confidence in science and technology show up in his notion of social reform. Like many thoughtful observers of the Depression, Dewey believed that capitalism was unable to manage itself. He was obviously right that government intervention and planning would be essential to the survival of modern society. But Dewey's sensible suggestions for reform are at times expressed in a rhetoric that sounds more technocratic than he may have intended. He writes in *Philosophy and Civilization*, for example, that "[s]cience has hardly been used to modify men's fundamental acts and attitudes in social matters," and, a few pages later, he recommends "scientific procedures for the control of human relationships and the direction of the social effects of our vast technological machinery."[34] In such passages he seems to call for the substitution of expert control of collective behavior for the market mechanism without leaving much room for a democratic alternative.

Now, this is not at all what I have in mind by the democratization of technology! Where are the popular movements, the suppressed needs of marginalized peoples, the insights of the laity ignored by a priestly cast of experts? All of this is acknowledged elsewhere by Dewey and Hickman, but somehow that does not prevent them from identifying "technology" and democracy on occasion. Apparently, Dewey did not have a strong a sense of the tension between science and technology and the everyday lifeworld in which meanings and significance are formed and democratic initiatives nurtured.

Technology as Lifeworld

As I read him Dewey's position on technology is ambiguous. In some texts he argues that technology is primarily a means to ends. In this, technology differs, he argues, from art. But at times he argues that the design of technology is the design of the social world itself, and not just instrumental to extrinsic ends. These passages hint at something like the synthesis of Heidegger's philosophies of art and technology to which I referred above. In *Philosophy and Civilization* he writes that "science has established itself and has created a new social environment," an environment, he argues, which must be improved.[35] There is a passage in *Art as Experience* where he remarks that the work of shaping and designing everyday objects tends to suit the needs of the whole self rather than a narrow efficiency of purpose. "All that we can say is that in the absence of disturbing contexts, such as production of objects for a maximum of private profit, a balance tends to be struck so that objects will be satisfactory—'useful' in the strict sense—to the self as a whole, even though some specific efficiency be sacrificed in the process."[36]

This is a possible bridge between my position and Dewey's. Critical theory of technology analyzes technology as a life environment subject to interpretation and judgment according to a wide range of norms and not merely in technical terms. So long as technology is viewed primarily as a means to an end, its political significance will be overlooked. As can be seen from these quotations, Dewey does recognize that capitalism distorts technology and the society it shapes, and he believes that technical work tends naturally to address the wider range of needs of the self. The instrumentalization theory helps to formulate a similar argument more precisely as a philosophy of technology and as a basis for evaluating specific designs.

It is of course easier today to go beyond Dewey's ambiguous statement of the case than fifty or more years ago. We have seen the rise of various public movements for democratic control of technology such as the environmental movement. These movements give us a hint of how publics can recognize themselves and defend their common interests even under the new conditions created by modern technology.

Democracy thus finds a new rationale and a new mission in technologically advanced societies. Up to now, democracy has been conceived primarily as a limit on the state. Those who live under the laws should make the laws. This is common sense. But today democracy must also protect us from the unintended consequences of technical action. Those threatened by technology must control technology. This idea is much more difficult to grasp. It defies the short-term logic of both the market and the electoral system. It requires a new common sense informed by scientific knowledge of nature and health. Dewey was perhaps the first philosopher to appreciate the task in a general way. On this score we can find in him a predecessor of note.

Conclusion

Philosophy of technology has come a long way since Heidegger and Dewey. Inspiring as are these thinkers, we need to devise our own response to the situation in which we find ourselves. Capitalism has survived its various crises and now organizes the entire globe in a fantastic web of connections with contradictory consequences. Manufacturing flows out of the advanced countries to the low wage periphery as diseases flow in. The Internet opens fantastic new opportunities for human communication, and is inundated with commercialism. Human rights proves a challenge to regressive customs in some countries while providing alibis for new imperialist ventures in others. Environmental awareness has never been greater, yet nothing much is done to address looming disasters such as global warming. Nuclear proliferation is finally fought with energy in a world in which more and more countries have good reasons for acquiring nuclear weapons.

It is sad that for the most part philosophy has nothing to contribute to the discussion of this astonishing situation. Philosophy of technology is marginalized in the profession with the result that philosophy itself has become ever more marginal to the culture. Unfortunately, for reasons social scientists ought to examine, most discussion of technology in the social sciences is politically toothless. I am gratified that our debate breaks the rules and addresses the philosophical and political implications of real world issues. The importance of our discussion here cannot therefore be measured in professional terms.

Building an integrated and unified picture of our world has become far more difficult as technical advances break down the barriers between spheres of activity to which the division between disciplines corresponds. I believe that critical theory of technology offers a platform for reconciling many apparently conflicting strands of reflection on technology. Only through an approach that is both critical and empirically oriented is it possible to make sense of what is going on around us now. The first generation of Critical Theorists called for just such a synthesis of theoretical and empirical approaches.

Critical Theory was above all dedicated to interpreting the world in the light of its potentialities. Those potentialities are identified through serious study of what is. Empirical research can thus be more than a mere gathering of facts and can inform an argument with our times. Philosophy of technology can join together the two extremes—potentiality and actuality—norms and facts—in a way no other discipline can rival. It must challenge the disciplinary prejudices that confine research and study in narrow channels and open perspectives on the future.

Notes

1. I have responded to several of my critics at greater length in separately published debates. Here are the references to my responses: "Constructivism and Technology Critique: Response to Critics," *Inquiry* (Summer 2000): 225–38; "Do We Need a Critical Theory of Technology? Reply to Tyler Veak," *Science, Technology, and Human Values* (Spring 2000): 238–42; "Will the Real Posthuman Please Stand Up! A Response to Fernando Elichirigoity," *Social Studies of Science* 30, no. 1 (February 2000): 151–57; "The Ontic and the Ontological in Heidegger's Philosophy of Technology: Response to Thomson," *Inquiry* 43 (Dec. 2000): 445–50; "Technical Codes, Interests, and Rights: Response to Doppelt," *The Journal of Ethics* 5, no. 2 (2001): 177–95; "Pragmatism and Critical Theory," *Techné* 7, no. 1 (Fall 2003): 42–48. The criticisms and these responses are available online at http://www.sfu.ca/~andrewf/symposia.html.
2. Herbert Marcuse, *One-Dimensional Man* (Boston: Beacon Press, 1964), xlvii.
3. Andrew Feenberg, *Heidegger and Marcuse: The Catastrophe and Redemption of Technology* (New York: Routledge, 2004).
4. Andrew Feenberg, *Critical Theory of Technology* (New York: Oxford University Press, 1991).
5. Andrew Feenberg, "Building a Global Network: The WBSI Experience," in *Global Networks: Computerizing the International Community*, ed. L. Harasim (Cambridge: MIT Press, 1993), 185–97.
6. Herbert Marcuse, "Beiträge zu einer Phänomenologie des Historischen Materialismus," in *Herbert Marcuse Schriften: Band I* (Frankfurt: Suhrkamp Verlag, 1978), 364–65.
7. For a discussion of this "empirical turn," see Hans Achterhuis, "Andrew Feenberg: Farewell to Dystopia," in *American Philosophy of Technology*, ed.

H. Achterhuis (Bloomington and Indianapolis: Indiana University Press, 2001.)

8. See Andrew Feenberg, *Alternative Modernity: The Technical Turn in Philosophy and Social Theory* (Los Angeles: University of California Press, 1995); Andrew Feenberg, *Questioning Technolog.*(London and New York: Routledge, 1999); Andrew Feenberg, *Transforming Technology: A Critical Theory Revisited* (New York: Oxford University Press, 2002).
9. The implied reference is to the concept of a godlike "view from nowhere." If it were not too cute, one might rephrase the point here as a "do from knowhere," i.e., action understood as just as indifferent to its objects as detached knowing.
10. Edward Tenner, *Why Things Bite Back: Technology and the Revenge of Unintended Consequences* (New York: Alfred A. Knopf, 1996).
11. Michel de Certeau, *L'Invention du Quotidien* (Paris: UGE, 1980).
12. Andrew Feenberg, "Modernity Theory and Technology Studies: Reflections on Bridging the Gap," in *Modernity and Technology*, ed. T. Misa, P. Brey, and A. Feenberg (Cambridge: MIT Press, 2003).
13. Stump, this volume, 6.
14. Wiebe Bijker, Thomas P. Huges, and Trevor Pinch, *The Social Construction of Technological Systems* (Cambridge: MIT Press, 1987); Wiebe Bijker and John Law, eds., *Shaping Technology/Building Society: Studies in Sociotechnical Change* (Cambridge: MIT Press, 1992).
15. Andrew Feenberg and Darin Barney, eds., *Community in the Digital Age: Philosophy and Practice* (Lanham: Rowman and Littlefield, 2004).
16. Glazebrook, this volume, 42.
17. Glazebrook, this volume, 46.
18. Hubert Dreyfus, "Heidegger on Gaining a Free Relation to Technology," in *Technology and the Politics of Knowledge*, ed. Andrew Feenberg and Alastair Hannay (Bloomington and Indianapolis: Indiana University Press, 1995), 102–103.
19. Martin Heidegger, *The Question Concerning Technology*, trans. W. Lovitt (New York: Harper and Row, 1977), 30.
20. Martin Heidegger, "Traditional Language and Technological Language," trans. W. Gregory, in *Journal of Philosophical Research* XXIII (1998): 40–41.
21. Ibid.
22. Gerald Doppelt, "Rawls' System of Justice: A Critique from the Left," *NOUS* 15, no. 3 (1981).
23. See Hans Achterhuis, ed., *American Philosophy of Technology* (Bloomington: Indiana University Press, 2001).
24. Borgmann, this volume, 101.
25. Albert Borgmann, *Technology and the Character of Contemporary Life* (Chicago: University of Chicago Press, 1984).
26. Borgmann, this volume, 107.
27. Thompson, this volume, 128.
28. Thompson, this volume, 131.
29. Feenberg, *Questioning*, chapter 3.
30. Light, this volume, 149.
31. For more on Light's position, see Andrew Light, "Restoring Ecological Citizenship," in *Democracy and the Claims of Nature*, ed. B. Minteer and B. P. Taylor (Lanham: Rowman and Littlefield, 2002), 153–72.

32. Larry Hickman, *Philosophical Tools for Technological Culture* (Bloomington and Indianapolis: Indiana University Press, 2001), 80.
33. John Dewey, *Individualism Old and New* (New York: Capricorn Books, 1962), 30, 98.
34. John Dewey, "Science and Society," in *The Philosophy of John Dewey*, ed. J. McDermott (Chicago: University of Chicago Press, 1981), 393, 397.
35. Ibid., 392.
36. John Dewey, *Art as Experience* (New York: Perigree, 1934), 115.

Contributors

Albert Borgmann is Regents Professor of Philosophy at the University of Montana. His area of specialization is the philosophy of society and culture with particular emphasis on technology. Among his publications are *Technology and the Character of Contemporary Life* (University of Chicago Press, 1984), *Crossing the Postmodern Divide* (University of Chicago Press, 1992), *Holding on to Reality: The Nature of Information at the Turn of the Millennium* (University of Chicago Press, 1999), and *Power Failure: Christianity in the Culture of Technology* (Brazos Press, 2003).

Simon Cooper is a Lecturer in Mass Communications and Writing at Monash University, Gippsland. He is a member of the Gippsland-based Research Unit for Work and Communications Futures, and a member of the HUMCASS-based Active Citizens/New technologies research team. His current research interests are critical theory, psychoanalysis, politics, civics, film, and cybercultures. In addition to these more general areas of research, Simon has completed more specific research within the local region. In 1998 he completed a social impact study on the introduction of e-commerce in the Latrobe region for the Latrobe Shire. Simon is an editor of *Arena* journal and is a regular contributor to *Arena* magazine. His publications include *Technoculture and Critical Theory* (Routledge, 2002).

Gerald Doppelt is a Professor of Philosophy at the University of California at San Diego. He has also taught at the University of Pennsylvania, and as a visiting lecturer at U.C. Berkeley and the University of Illinois, Chicago Circle. His two main areas of philosophical interest are the philosophy of science and political theory. In his Ph.D. thesis and subsequent research, Doppelt has been concerned with the conflicts between empiricist, historicist, and pragmatic conceptions of science (including social science), especially concerning the role of observational data in validation. Doppelt's research interests in political theory focus on developing a philosophical dialogue between liberal and radical conceptions of social justice and social theory itself. This also involves an interest in explicating Marxism as a distinctive approach to philosophical problems such as that of fact-value, knowledge-ideology, and normative ethics. He has published numerous peer-reviewed articles and book chapters on these topics.

Andrew Feenberg is a Canada Research Chair in Philosophy of Technology in the School of Communication, Simon Fraser University. He has also taught for many years in the Philosophy Department at San Diego State University, and at Duke University, the State University of New York at Buffalo, the Universities of California, San Diego and Irvine, the Sorbonne, the University of Paris-Dauphine, the Ecole des Hautes Etudes en Sciences Sociales, and the University of Tokyo. He is the author of *Lukacs, Marx and the Sources of Critical Theory* (Rowman and Littlefield, 1981; Oxford University Press, 1986), *Critical Theory of Technology* (Oxford University Press, 1991), *Alternative Modernity* (University of California Press, 1995), and *Questioning Technology* (Routledge, 1999). A second edition of *Critical Theory of Technology* has appeared with Oxford in 2002 under the title *Transforming Technology*, and he just recently published *Heidegger, Marcuse and Technology: The Catastrophe and Redemption of Enlightenment* (Routledge 2004). Translations of several of these books are available. Feenberg is also co-editor of *Marcuse Critical Theory and the Promise of Utopia* (Bergin and Garvey Press, 1988), *Technology and the Politics of Knowledge* (Indiana University Press, 1995), and *Modernity and Technology* (MIT Press, 2003), and *Community in the Digital Age* (Rowman and Littlefield, 2004). His co-authored book on the French May Events of 1968 appeared in 2001 with SUNY Press under the title *When Poetry Ruled the Streets*. In addition to his work on Critical Theory and philosophy of technology, Feenberg has published on the Japanese philosopher Nishida Kitaro. He is also recognized as an early innovator in the field of on-line education, a field he helped to create in 1982. He is currently working on the TextWeaver Project on improving software for on-line discussion forums under a grant from the Fund for the Improvement of Post-Secondary Education of the U.S. Department of Education.

Trish Glazebrook is an associate professor of philosophy at Dalhousie University and has also taught at the University of Toronto, Colgate University, and Syracuse University. A Board Director of the International Association of Environmental Philosophers, she is the author of *Heidegger's Philosophy of Science*

(Fordham University Press, 2000), and *Heidegger's Critique of Science* (ed.) (SUNY Press, forthcoming). She has also published various articles on Heidegger, ancient and modern science, environmentalism, and feminism.

Larry A. Hickman is a Professor of Philosophy and Director of The Center for Dewey Studies, Southern Illinois University at Carbondale. His most recent books include *Philosophical Tools for Technological Culture: Putting Pragmatism to Work* (Indiana University Press, 2001) and *John Dewey's Pragmatic Technology* (Indiana University Press, 1990). In addition, Hickman has edited numerous volumes on the work of John Dewey, including *The Influence of Darwin on Philosophy and Other Essays in Contemporary Thought by John Dewey: A Critical Edition* (Southern Illinois University Press, forthcoming), and *The Correspondence of John Dewey, Vol. I: 1871–1918* (InteLex Corporation, 1999).

Andrew Light is Associate Professor of Philosophy at the University of Washington. He holds a joint appointment with the Evans School of Public Affairs and an adjunct appointment with the Department of Geography. He is also a Research Fellow at the Institute for Environment, Philosophy & Public Policy at Lancaster University (U.K.), and a Faculty Fellow at the Center for Sustainable Development in the School of Architecture at the University of Texas at Austin. Prior to coming to NYU, Light was assistant professor of philosophy and environmental studies at SUNY Binghamton, and before that, assistant professor of philosophy at The University of Montana. Prior to Montana, Light held a three year Canadian Tri-Council postdoctoral research fellowship in the Environmental Health Program (School of Medicine) at the University of Alberta focusing on environmental risk management. His primary areas of interest are environmental ethics and policy, philosophy of technology, moral and political philosophy, and aesthetics. Light is the author of more than sixty articles and book chapters on these topics, and is editor or co-editor of fourteen books, including *The Aesthetics of Everyday Life* (Columbia University Press, 2005), *Animal Pragmatism: Rethinking Human-Nonhuman Relationships* (Indiana University Press, 2004), *Moral and Political Reasoning in Environmental Practice* (MIT Press, 2003), *Environmental Ethics: An Anthology* (Blackwell, 2003). He is also the author of a new book on philosophy and film, *Reel Arguments: Film, Philosophy, and Social Criticism* (Westview, 2003). He is currently completing a monograph on ethical issues in restoration ecology, tentatively titled *Restoring the Culture of Nature*. Funding for this project has been provided by an individual scholar award from the National Science Foundation and a Harrington Faculty Fellowship from the University of Texas at Austin. Future research will focus on urban environmental issues.

David J. Stump is a philosopher of science at the University of San Francisco. Active in the History of the Philosophy of Science Working Group (HOPOS), he has published on the philosophy of Henri Poincaré, naturalized philosophy of science, and science and society. He is co-editor with Peter Galison, of *The Disunity of Science*.

Paul B. Thompson is W. K. Kellogg Chair in Agricultural, Food and Community Ethics at Michigan State University in the Philosophy Department, with partial appointments in the Agricultural Economics and Resource Development Departments. Previously he held positions as Distinguished Professor of Philosophy and Director, Center for Food Animal Productivity and Well-Being, at Purdue University, and Professor of Philosophy and Agricultural Economics and Director, Center for Science and Technology Policy and Ethics, at Texas A&M University. His research interests include: American pragmatist approaches in practical ethics, environmental ethics, risks and ethics of agricultural and food biotechnology, science policy, philosophy of technology, and philosophy of economics. He is the author of *The Spirit of the Soil: Agriculture and Environmental Ethics; The Ethics of Aid and Trade; Food Biotechnology in Ethical Perspective*, and co-editor of *The Agrarian Roots of Pragmatism*.

Iain Thomson is an assistant professor of Philosophy at the University of New Mexico, and has also taught at Rice University and the University of California at San Diego. He has published articles on Heidegger in *Inquiry*; *The International Journal of Philosophical Studies*; *Philosophy Today*; and *Enculturation*, and is currently finishing a book on the development of Heidegger's thought, titled *The End of Ontotheology: Understanding Heidegger's Turn, Method, and Politics*.

Edward J. Woodhouse is an associate professor of political science, Department of Science and Technology Studies, Rensselaer Polytechnic Institute. Woodhouse works in the tradition of democratic theorists Robert Dahl and Charles Lindblom in conceptualizing a fairer and wiser political economy, with particular attention to the strategies, institutions, and processes for governing science and technology more democratically. His present research concerns: policy options for accelerating the adoption of Green Chemistry (benign chemical processes and products); media and interest-group shortcomings in promoting intelligent public discussion of agricultural biotechnology; design professionals' contributions to overconsumption in the affluent nations; and conceptualization of economic democracy as a parallel to governmental democracy. His books include: *Averting Catastrophe: Strategies for Regulating Risky Technologies* (University of California Press, 1986, with

Joseph Morone); *The Demise of Nuclear Energy?: Lessons for Democratic Control of Technology* (Yale University Press, 1989, with Joseph Morone), *The Policy-Making Process*, 3rd edition (Prentice-Hall, 1993, with Charles E. Lindblom). He is currently completing *Inventing Our Grandchildren's World: Envisioning a Commendable Technological Civilization*.

Index

Abbey, Edward, 137, 138, 141
Achterhuis, Hans, 105, 110, 208, 209
actor-network theory (or theorists), xx, 6, 25, 26, 80, 112, 125, 167, 177
Adbusters (magazine), 51
Adorno, Theodore, viii, xix, 73, 74, 75, 80
Advances in Social Theory and Methodology:, Toward Integration of Micro- and Macro-
Sociologies (book), 18
Advil, 159
AEC. *See* Atomic Energy Commission.
Aesthetic Dimension, The:, Toward a Critique of Marxist Aesthetics (book), xix
Africa, 140
A Friend of the Earth (book), 136, 138, 151
agency, 29, 35, 55, 76, 77, 92, 93, 94, 179, 182, 196, 197, 202
ahistoricism, xvi, 54, 55, 56, 57, 59, 65
Aim and Structure of Physical Therapy, The (book), 17
Alcoff, Linda, 52
Alternative Modernity:, The Technical Turn in Philosophy and Social Theory (book), xix,,
xx, 17, 36,, 42, 43, 51, 65, 99, 114, 115, 128, 130, 134, 156, 171, 178, 189, 201, 209
America. *See* United States of America.
American Declaration of Independence, 119
American Electric Power Company, 171
American Philosophy of Technology (book), 110, 208, 209
Americas, the, 122
Anastas, Paul T., 171
Anderson, Joel, 66, 71, 80, 81
A Nice Derangement of Epistemies:, Post Positivism in the Study of Science from Quine to Latour (book), 17
Animal Liberation Front, 141
Animal Machine (book), 52
anti-essentialism (or nonessentialism), xv, xvi, 8, 10, 12, 14, 41, 73, 74, 85, 86, 90
Apel, 72
Aramis or the Love of Technology (book), 17
Arena Journal, 35
Arena Magazine, 36
Arkwright, Sir Richard, 46
Art as Experience (book), 206, 210
A Social History of American Technology (book), 134
A Strategy of Decision (book), 172
Atlantic Monthly, The (newsmagazine), 108, 110
Atomic Energy Commission (AEC), 157
Auden, W.H., 65
autonomous hybrids, 14
Autonomous Technology:, Technics-out-of-control as a Theme in Political Thought (book), 171

Barnes and Noble, 106
Barney, Darin, 209
Baudrillard, Jean, xiv, 54, 57, 66
Beauvoir, Simone de, 41, 51
Beats, the, 175
Beckenbauer, Franz, 63
Benedict, J., 66
Bentham, Jeremy, 15
Berg, Paul, 27
Berlin, 11
Berlin, Isaiah, 198
Biehl, Janet, 51
Bijker, Wiebe, xx, 3, 7, 16, 113, 167, 188, 209
Bildung, 61
biological determinism, 41, 44
biopiracy, 24
biotechnology, 20, 21, 22, 23, 24, 26, 27, 28, 33, 34, 120, 189, 190
Bloor, David, xx, 6, 17, 187
Bodies in Technology (book), 134
Bookchin, Murray, 141, 143, 151
Books and Culture (journal), 110
Borgmann, Albert, vi, xvii, 54, 101, 112, 115, 116, 120, 132, 133, 134, 144, 152, 185, 187, 198, 199, 200, 201, 203, 209, 211
Boydston, Jo Ann, 81
Boyle, T. C., 136, 137, 138, 141, 151
Brave, Ralph, 27, 36
Braybrooke, David, 172
Breyman, Steve, 172
Brey, P., 209
Briffault, H., 17
Britain. *See* United Kingdom.
British Educational Research Journal, 36
Brittan, Gordon, 105, 110, 144
Brodsly, David, 69
Buchwald, Jed Z., 18

Caddick, Alison, 28, 33, 36
California, 140, 144, 204,
Callon, Michel, 13, 16, 18, 113
Cambridge Companion to Heidegger, The (book), 67
Campbell, Sue, 50
Capital:, A Critical Analysis of Capitalist Production (book), xx
capitalism (or capitalist economy), vii, viii, ix, xv, 4, 33, 48, 87, 91, 106, 118, 122, 128, 129, 143, 153, 155, 170, 180, 181, 185, 200, 205, 206, 207
Capuzzi, Frank, 51, 67
Carter, Alan, 110
Casper, Monica, 171
Catholic Church, the, 43
Cheney, George, 110
Chicago Board of Trade, 125, 127
Chicago Wilderness, 149
Christ, Carol, 47, 52
Cicourel, Aaron, 18,
Cixous, Helen, 45, 51
class, 26
cloning, 26, 29
Cold War, 157
Columbia University 78
commodification, xvii, 112, 115, 116, 117, 118, 119, 120, 121, 122, 123, 125, 126, 127, 128, 129, 133, 201–202; and secondary rationalization, 112–135; structural, xvii, 112, 117, 118,, 121, 122, 127, 201; technological, xvii, 115, 116, 117, 118, 120, 121, 122, 123, 125, 126, 128, 129, 131, 132, 133, 201. *See also* technology.
commodity fetishism, xiv
commodity market(s), 126
Commoner, Barry, 138, 139, 140, 142, 143, 144, 148, 151, 202
Community in the Digital Age:, Philosophy and Practice (book), 209
communicative action, x
communicative reason, 4
Computers in the Human Context:, Information Technology, Productivity, and People
(book), 81
constructivism, 25, 54, 168, 178, 180, 184, 185, 186, 187, 188, 191. *See also* social
constructivism, *and* Social Construction of Technology (SCOT).
contextualism (or contextualist historians of technology), xx, 79
Contributions to Philosophy (From Enowning) (book), 66, 68, 69
Cooper, Simon, v, xv, 19, 35, 36, 185, 187, 188, 189, 190, 191, 192, 211

Copernican Revolution, the, 16
Corea, Gena, 50
Counterrevolution and Revolt (book), xix
Cowan, Ruth Schwartz, 134
Crease, Robert P., 110
critical theory (or theorists), iii, xiv, 55, 56, 71, 72, 73, 74, 75, 76, 78, 80, 128, 177, 178, 193, 205, 208. *See also* Feenberg.
Critical Theory, Marxism, and Modernity (book), xix
Critical Theory of Technology (book), xix, 65, 99, 177, 178, 208
Criticism and the Growth of Knowledge (book), 52
Cronon, William J., 127, 134
culture, 26
Culture Jam (book), 51
cultural theory, xiv
Customs in Common:, Studies in Traditional and Popular Culture (book), 134
Cutting into the Meatpacking Line:, Workers and Change in the Rural Midwest (book), 134
cyborg (ontology) 24, 25, 27

Darwin, Charles, 205
Darwinian Revolution, 16
Dawson, Frank G., 171
d'Eaubonne, Francoise, 47, 51
Death of Nature, The:, Women, Ecology, and the Scientific Revolution (book), 52
de Certeau, Michel, 183, 209
decommodification, xvii, 123, 126, 127, 129, 130, 199, 201, 202
deep ecology (or deep ecologists), 141, 143. *See also* social ecology.
Demise of Nuclear Energy, The?:, Lessons for Intelligent Democratic Control of Technology (book), 171
democracy, xvi, 43, 44, 48, 50, 55, 72, 73, 78, 80, 85, 94, 98, 136, 145, 147, 148, 149, 168, 202, 203; and technology, iii, 48, 50, 55, 61, 62, 72, 73, 78, 85, 94, 98, 108, 136–152, 157, 177, 183, 185, 196, 206. *See also* Feenberg.
Democracy and Technology (book), 171
Democracy and the Claims of Nature (book), 209
Democratic Justice (book), 99
democratic rationalization, xii, 80, 97, 107, 139, 147, 150
Democratizing Technology (book), 36
deoxyribonucleic acid (DNA), 121
Department of Defense (U.S.), 104, 109
Depression, the, 205
Derrida, Jacques, 58
design. *See* technology, design of.
design choice, xi, xii, xiv, 78
design process, xvii, 62, 78
deskilling, 125, 126, 203
Dewey, John, xvi, 72, 74, 75, 76, 77, 78, 79, 80, 81, 181, 204–207, 210
Dialectic of Enlightenment (book), xix
dialectic of technology. *See* Feenberg.
Diamond, Irene, 52
Discipline and Punish:, The Birth of the Prison (book), 13, 18
Disclosing New Worlds: Entrepreneurship, Democratic Action, and the Cultivation of Solidarity (book), 68, 69
Discourse on Thinking (book), 66, 69
Disturbing the Universe (book), 171
Disunity of Science:, Boundaries, Contexts, and Power, 17
DNA. *See* deoxyribonucleic acid.
domination, logic of, 37, 47, 48
Doppelt, Gerald, v, xvi, xvii, 70, 85, 182, 196, 197, 198, 208, 209, 211
Dow Chemicals, 159
Dowie, Mark, 140, 143, 144, 145, 151, 152
Draize test, the. *See* LD50 test.
Dreyfus, Hubert L., 60, 61, 64, 67, 68, 69, 70, 192, 193, 209
Duhem, Pierre, 17
Duhem-Quine thesis, 6
Dyson, Freeman, 171

Earth First!, 140, 141, 142, 144
Earth Liberation Front, 141
ecofeminism (or ecofeminists), xvi, 37, 39, 40, 41, 43, 47, 48, 191. *See also* feminism.
Ecological Feminism (book), 52
Economy and Society:, An Outline of Interpretive Sociology (book), xix

EDF. *See* Environmental Defense Fund.
Edison, Thomas, 10
1844 Manuscripts (book), 115, 119
Ehrlich-Commoner debate over population: xviii, 138, 139–145. *See also* Ehrlich, *and*
Commoner.
Ehrlich, Paul, 138, 139, 140, 143, 144, 146, 147
Ellul, Jacques, 28, 54, 73, 77, 142, 144, 187
Emad, P., 66, 69
emancipation, ix, x, 33, 71, 205
Empire (book), 25, 35
Encounters and Dialogues with Martin Heidegger:, 1929–1976 (book), 69
Enculturation (journal), 69
Energy Possibilities (book), 110
enframing:, 45, 57, 58, 59, 62, 193, 194, 195. *See also* Heidegger, *and* technology.
Engels, Frederick, xx
Enlightenment, the, viii, 4, 9, 14, 16, 205
Environmental and Natural Resource Economics (book), 172
environmental change, vi, 136–152
Environmental Defense Fund (EDF), 143, 144, 164, 169
Environmental Ethics (journal), 50, 52
environmentalism (or environmental movement), 37, 44, 102, 138, 139, 140, 141, 143, 144, 145, 148, 150, 176, 178, 193, 201, 202, 203
environmental materialists, 143
environmental ontologists, 143
Epistemology and Environmental Philosophy (book), 50
essentialism (or essentialist philosophy of technology), v, xiv, xv, xvi, xx, 4, 5, 8, 9, 12, 13, 14, 15, 20, 29, 37, 41, 53, 54, 55, 56, 59, 65, 72, 73, 75, 87, 88, 90, 91, 102, 113, 185,, 186, 187, 188, 189, 191
Ethics (journal), 99
Europe, 122
experimental medicine, 95, 97

FAIR. *See* Federation for American Immigration Reform.
fallibilism, 79
false consiousness, viii
Federal Reserve. *See* United States Federal Reserve.
Federation for American Immigration Reform (FAIR), 140
Feenberg, Andrew:, and AIDS (or AIDS activism or medicine), xiii, xvii, xviii, 20, 41, 42, 43, 86, 89, 93, 95, 96, 97, 103, 104, 114, 128, 129, 130, 140, 156, 176, 201, 203; and (alternative) modernity, vii, xvi, xviii, 22, 24, 27, 42, 48, 50, 59, 66, 67, 68, 87, 88, 98, 153, 155, 167; and cyborg ontology, 24, 25; and democratic technology (or democracy), vii, xvii, 55, 61, 62, 73, 78, 87, 88, 90–91, 92, 95, 96, 104, 105, 108, 109, 110, 132, 147, 148, 149, 150, 151, 154, 155, 160, 162, 177, 182, 185, 201, 203, 204, 206, 207; and democratic or subversive rationality (or rationalization) xi, xii, xiii, xvii, 40, 41, 42, 47, 48, 80, 145, 147, 150, 155; and dialectic of technology, 32, 33, 93, 114; and ecofeminism (or ecofeminists or feminism), v, 37–52, 191; and environmentalism, xviii, 136–152; and essentialism (or non-essentialism), v, xv, xx, 3, 8, 12, 14, 15, 20, 23, 24, 26, 29, 34, 42, 44, 45, 46, 53–70, 72, 73, 74, 80, 85, 86, 87, 88, 90, 91, 97, 177, 186, 187, 204; and fatalism, 28, 55, 56, 59, 60, 67; and governance of technology, 154, 155; and his critical theory (or philosophy) of technology, iii, vii, viii, x–xii, xvi, 5, 19, 20, 22, 46, 53, 98, 110, 112, 113, 115, 131, 168, 178, 198, 200–201, 208; and his demystification of technology, 86, 87, 91, 98, 196; and his progress from critical theory to pragmatism, v, 71–81; and his theory of instrumentalization, xiii–xiv, xvi, xvii, xx, 8, 9, 21, 22, 26, 29, 33, 45, 103, 108, 109, 128, 130, 132, 185–187, 189, 190, 191, 192, 202, 206; and his theory of technological transformation, vii, xv, xvi; and invigorated (or critical)

constructivism, 74, 75, 78,, 178; and neutrality of technology, 182; and operational autonomy, 180–182, 185, 203, 204; and posthumanism, xv–xvi, 19–36, 188–189; and reform of technology, v, xv, xvii, xviii, 87, 101–111, 114, 133, 138, 139, 145, 146, 148, 149, 150, 151, 154, 201; and Social Construction of Technology (SCOT), xi, xv, xx, 4, 167; and social constructivism, 6, 7, 8, 13, 25, 42, 45, 54, 66, 72, 80, 103, 154, 168, 177, 184, 185, 186, 187, 188, 191; and technical code, vii, xii, 42, 43, 44, 46, 47, 48, 49, 50, 86, 88, 90, 91, 95–97, 185, 201, 208; and technological concretization, 201; and technological consciousness, xiv, xv; and the principle of 'conservation of hierarchy', xiii; and substantive conceptions of technology, 24, 35, 55, 56, 59, 61, 65, 72, 76, 187, 188, 189, 190, 196; and the concept of "participant interests", 87–91; and reflexivity, 21, 26, 28, 35; and technical action, 179–180, 189, 193, 207; and technological activism, 196; and technological (or technical) design, 42, 74, 78, 86, 88, 91, 96, 98, 113, 129, 177, 180, 184, 196, 203; and technological determinism (or determinists), 23, 26, 42,, 46, 74, 77, 103, 130, 131, 142, 156, 177, 184, 204; and technological innovation, 153, 154, 155, 156; and technological malleability, 153–170, 203
feminism (or feminist movement, feminist philosophy, feminists), xvi, 39, 41, 42, 44, 45, 46, 48, 49, 93, 102, 176, 192, 200
Feminist Epistemologies (book), 52
Feminist Theory:, From Margin to Center (book), 52
FDA. *See* United States Food and Drug Administration.
Fink, Deborah, 134
First National People of Color Environmental Leadership Summit (Oct. 1991), 145
Fischer, Michael, 145
Flores, Fernando, 67, 68, 69
Foreman, Dave, 141
Forester, Tom, 81
Fortune 500, the (corporate ranking system), 164
Foucault, Michel, xiv, 13, 15, 20, 29, 30, 36, 46, 49, 77, 183
Foucault Reader, The (book), 36
Four Arguments for the Elimination of Television (book), 51
France, 43, 106, 157, 176
Frankfurt Institute, xix
Frankfurt School (of critical theory) vii, x, xi, xii, xiii, xix, xx, 3, 19, 71, 80, 103, 176, 177, 179, 205
French Minitel system, 43, 93, 103, 114, 129, 130, 201, 203
French Teletel system, 86, 89
Freund, E., 66
Friedman, Milton, 172
Fromm, Erich, xix
Fukuyama, Francis, 23
Fuller, Steve, 31, 36
Future of Ideas, The (book), 110

Galison, Peter, 17
gay rights, 43
gender (constructions), 41, 42, 44, 45, 46
General Accounting Office. *See* United States General Accounting Office.
General Electric, 156, 157
German Fundis, 140, 142
German Green Party, 140, 142
German Realos, 140
Gerth, H., 18
Gesamtkunstwerk, 66. *See also* Nietzsche.
Gilligan, Carol, 45, 51
Glazebrook, Trish, v, xvi, 37, 50, 66, 185, 187, 191, 192, 193, 194, 209, 212
Glenn, Cathy B., 52
globalization, 22, 26, 72, 139
Global Networks:, Computerizing the International Community (book), 208
GNU/Linux. *See* Linux.
Gogh, Van, 69
green chemistry, xviii, 153, 158–162, 170, 171

Green Chemistry:, Theory and Practice (book), 171
Gregory, W., 66, 209
Griffin, Susan, 47, 51
Guignon, C., 67

Habermas, Jürgen, ix, x, xi, xiii, xix, 4, 16, 54, 55, 61, 71, 72, 73, 80, 113, 114, 176
Hair, Jay, 143
Hamlett, Patrick, 172
Hannay, Alastair, 50, 68, 81, 99, 134, 171, 209
Harasim, L., 208
Haraway, Donna, 19, 24, 29
Harding, Sandra, 52
Hardt, M., 25, 35
Harries, L., 69
Harris, Jonathan M., 172
Harrison, Ruth, 52
Hart, J. G., 69
Hegel, Georg Wilhelm Friedrich, 23, 64, 69, 102, 194, 199, 205
Heidegger and Marcuse:, The Catastrophe and Redemption of Technology (book), 208,
Heidegger, Martin, xvi, xx, 4, 9, 16, 19, 23, 27, 28, 33, 36, 37, 38, 39, 40, 41, 45, 46, 47, 50, 51, 53, 54, 55, 56, 57, 58, 59, 60, 61, 62, 63, 64, 65, 66, 67, 68, 69, 80, 113, 133, 142, 144, 155, 175, 176, 177, 178, 179, 180, 183, 186, 187, 188, 192, 193, 194, 195, 205, 207, 208, 209
Heidegger, Philosophy, Nazism (book), 67
Heim, M., 69
Held, David, xix
Herbert Marcuse Schriften:, Band I, 208
Hermann, F.-W. von, 66
hermeneutics, xiv
Hess, David, 172
Hickman, Larry A., v, xvi, 71, 101, 110, 112, 132, 135, 204, 205, 206, 210, 213
Higgs, E., 134, 135, 152
Hinkson, John, 31, 36
Hisschemöller, Matthijs, 173
Hofstadter, A., 69
Holding on to Reality:, The Nature of Information at the Turn of the Millennium (book), 134
Holland, Nancy, 50
Holmes, Helen B., 50
Hooks, Bell, 52
Horkheimer, Max, viii, xix, 73, 74, 75, 80
Hoskins, Betty B., 50
How Experiments End (book), 17
Hrachovec, Herbert, 108, 109, 110
Hughes, Thomas P., xx, 3, 7, 10, 11, 14, 16, 17, 18, 188, 209
human-technological network(s), 25. *See also* cyborg ontology.
Hume, David, 13
Huntington, Patricia, 50

ibuprofen, 159
Ihde, Don, 74, 112, 118, 119, 125, 134
In a Different Voice:, Psychological Theory and Women's Development (book), 51
Individualism Old and New (book), 205, 210
industrialization, 202
informationalism, 31
information revolution, 31
information society, 32, 202
Inquiry (journal), 16, 18, 50, 65, 67, 68, 99, 134, 208
Institute for Social Ecology, 141
Institutions, Institutional Change, and Economic Performance (book), 134
Intellectual History Newsletter, 80
International Association for Environmental Philosophy, 52
International Journal of Philosophical Studies, 66
instrumentality (or instrumentalism), 77, 78, 79, 80
instrumentalization (or instrumentalization theory), xiii, xiv, xv, xvii, 8, 9, 22, 29, 33, 45, 46, 103, 105, 107, 108, 109, 114, 119, 128, 130, 131, 132, 184, 186, 189, 190, 191, 196, 200
Internet, the, xiii, xvii, 30, 62, 63, 65, 103, 104, 124, 125, 129, 156, 177, 190, 195, 202, 207

Introduction to Critical Theory:, Horkheimer to Habermas (book), xix
Inventing America:, Jefferson's Declaration of Independence (book), 134
in-vitro fertilization (IVF), 21, 22, 27, 34
Ister, the, 64
IVF. *See* in-vitro fertilization.

Jackson, Michael, 136
Jacobs, Alan, 110
James, P., 35
Japanese Go, 20
Jefferson, Thomas, 119
Joas, Hans, 72
Johnson, Steven, 23, 35
Journal of Applied Philosophy, 110
Journal of Ethics, The, 208
Journal of Philosophical Research, 66, 209
Journal of Social Philosophy, 152
Jung, Hwa Yol, 38, 39

Kellner, Douglas, xix, 103
Kettering, E., 69
Klee, Paul, 62
Knorr-Cetina, Karin, 18
Knowledge and Social Imagery (book), xx
Knowledge, Power, and Participation in Environmental Policy Analysis (book), 173
knowledge society, 31
Korsch, Karl, xx
Kranzberg, Melvin, 74, 81
Krell, Farrell, 51, 67
Kripke, Saul, 65
"L"
L. A. Freeway:, An Appreciative Essay (book), 69
Laird, F. N., 147, 152
Lakatos, Imre, 49, 52
L'Arc (journal), 51
Lasn, Kalle, 46, 47, 51
Later Works, The:, 1925–1953 (book), 81
Latour, Bruno, xx, 3, 6, 7, 13, 14, 16, 17, 18, 19, 24, 25, 26, 29, 54, 112, 113, 115, 117, 118, 119, 121, 123, 125, 127, 130, 131, 133, 134, 201, 202
Law, John, xx, 113, 209
LD50 test (or the Draize test), 49
Le Bourgeois Gentilhomme (play), 5, 16
Le Féminism ou La Mort (book), 51
Left Green Network, 141
Lessig, Lawrence, 104, 110
Leviathian and the Air Pump (book), 18
liberalism, 44, 45
Light, Andrew, vi, xviii, 134, 135, 136, 152, 199, 202, 203, 204, 209, 213
Lindblom, Charles E., 172
Linux (or GNU/Linux), 107, 108, 109, 110, 111
L'invention du Quotidien (book), 209
Locke, John, 87, 90, 91, 94
London, 11
Losing Ground:, American Environmentalism at the Close of the Twentieth Century (book), 145, 151
Lowenthal, Leo, xix
Lovitt, William, 46, 66, 209
Luddite, 37, 64
Lukács, Georg, xx

MacKenzie, Donald A., xx
Maly, K., 66, 69
Malthus, 146
Man and World (journal), 70
Maraldo, J. C., 69
Marcuse, Herbert, vii, viii, ix, x, xix, 33, 53, 61, 66, 71, 73, 75, 76, 77, 81, 113, 114, 130, 175, 176, 177, 178, 179, 180, 186, 208
Market System, The:, What It is, How it Works, and What to Make of it (book), 172
Martin Heidegger and National Socialism (book), 69
Mander, Jerry, 46, 51
Man-Made Women (book), 50
Mann, Charles C., 108, 110
Mars, 16, 18
Martin, Brian, 172
Marxism, xi, xx, 71, 77, 101, 115, 127, 134, 175
Marx, Karl, xiv, xx, 55, 87, 115, 117, 118, 119, 133, 180, 181
Max Weber:, Essays in Sociology (book), 18
McCarthy, Thomas, xix

McCay, David S., 18
McDermott, J., 210
McLuhan, Marshall, 74
McMullin, E., 17
McNeill, William, 69
Mead, George Herbert, 72
media, viii, 31, 32
Media Foundation (Canada), 51
medical-industrial complex, 42, 43
Merchant, Carolyn, 47, 52
Metaphysical Foundation of Logic, The (book), 69
Miller, A.V., 69
Mills, C.W., 18
Minteer, B., 209
Miracles Under the Oaks (book), 152
Misa, T., 209
Missoula, Montana, 106
Missoulian (newspaper), 110
Mitcham, Carl, 170
modernity, vii, xv, 3, 4, 9, 11, 21, 32, 37, 50, 67, 68, 186, 198; alternative, vii, xvi, xviii, 19, 20, 26, 42, 48, 88, 89, 94, 98, 167, 177, 191; as the construction of technological systems, 3–18, 21. *See also* Weber.
Modernity and Technology (book), 209
modern individual, 32
modern society, vii, 14, 85, 98, 183; technological designs of, vii
modernization, vii, xiii, 138
Molière, 5, 17
Mondragon Cooperative Corporation, Spain, 106
Monthly Review (journal), 111
Morone, Joseph G., 171
Morse code, 194
Motrin, 159
MP3, 107, 108, 123
Mumford, Lewis, 179
Musgrave, Alan, 52

Naess, Arne, 143, 144
Naming and Necessity (book), 65
Napster, 123
National Audubon Society, 140
National Wildlife Federation (NWF), 143
Nation-Formation (book), 35
Nation, The (newsmagazine), 27, 36
nature, ix, x, 32
Natural Resources Defense Council, 140,
Nature's Metropolis:, Chicago and the Great West (book), 127, 134
Negations:, Essays in Critical Theory (book), xix
Negri, A., 25, 35
neo-Luddism:, 142, 145. *See also* Luddite.
Neske, G., 69
network(s), 26, 27, 29, 30, 33, 67, 93, 113
Networks of Power:, Electrification in Western Society, 18
New Woman/New Earth:, Sexist Ideologies and Human Liberation (book), 51
New York, 11, 102
New York Times, The (newspaper), 110, 111
Nietzsche, Frederick, 51, 57, 58, 67
Nietzschean ontotheology, 57, 68
Nietzsche, Volume Four:, Nihilism (book), 67, 69
Nietzsche, Volume 3:, The Will to Power as Knowledge and as Metaphysics (book), 51
Nieusma, Dean A., 173
Noble, David, 109, 111
Noddings, Nel, 45, 51
North America, 38
North, Douglas, 112, 134
NOUS (journal), 209
Nuclear Power:, Development and Management of a Technology (book), 171

OECD. *See* Organisation for Economic Cooperation and Development.
Oil, Chemical, and Atomic Workers, 140
one-dimensionalism (or one-dimensional society) viii, x, xi, xvi, 54, 56, 62, 65, 66, 176, 178, 180
One-Dimensional Man:, Studies in the Ideology of Advanced Industrial Society (book), xix, 186, 208
O'Neill, John, 110
On Heidegger's Nazism and Philosophy (book), 52
ontic ambivalence, 27

ontic *versus* ontological, 27, 193, 194, 195
ontological contradiction, 22, 23, 34
operationalism, 25
Ophels, William, 146
Orem, Frank, 140
Orenstein, Gloria Feman, 52
Organisation for Economic Cooperation and Development (OECD), 49
Our Posthuman Future (book), 23

P&G. *See* Procter and Gamble.
paideia, 61
Pandora's Poison:, Chlorine, Health, and a New Environmental Strategy (book), 172
Pasteur, Louis 14, 17
pasteurization, 113
Pasteurization of France, The (book), xx, 17
Pastore, John O., 171
Pathmarks (book), 69
patriarchy, 44
peace politics:, 48. *See also* Warren.
Peirce, C. S., 72
PERC. *See* perchlorethylene.
perchlorethylene (PERC), 161, 169
perspectivism, 79
Petzet, Heinrich W., 69
phenomenology, 39, 63
Phenomenology of Spirit (book), 69
Philosophy and Civilization (book), 205, 206
Philosophy and Geography (journal), 110
Philosophy of John Dewey, The (book), 210
Philosophical Review (journal), 18
Philosophical Tools for Technological Change (book), 110, 135, 204, 210
Pickering, Andrew, 7, 18
Pinch, Trevor, xx, 167, 188, 209
Pippin, Robert, 99
Plato (or Platonists or Platonism), 56, 62, 68
Poetry, Language, Thought (book), 69
Policy-Making Process, The (book), 172
Pollock, Friedrich, xix
posthumanism (or posthuman theory/world), xv, xvi, 19, 20, 24, 25, 26, 27, 28, 32, 35, 188–189. *See also* Feenberg.
postmodernism (or postmoderns):, 25, 58. *See also* postmodernity.
postmodernity:, 21, 32, 67. *See also* postmodernism.
Potter, Elizabeth, 52
pragmatism, 71, 72, 74, 75
Presidential Green Chemistry Challenge Awards Program:, Summary of 1996 Award
Entries and Recipients (microform), 171
Preston, Christopher, 50
Proceedings of the Aristotelian Society, 17
Procter and Gamble (P&G), 166, 167
property (rights), notion(s) of, 34, 87, 90–91, 121, 122, 123, 127, 128, 196. *See also* Locke.
Property, Power, and Public Choice:, An Essay on Law and Economics (book), 134
Prozac, 23
Public and its Problems, The (book), 78, 79

Question Concerning Technology and Other Essays, The (book), 50, 52, 56, 66, 67, 68, 69, 209
Questioning Technology (book), xix, xx, 8, 18, 35, 36, 42, 43, 45, 50, 51, 52, 53, 58, 65, 66, 68, 69, 71, 72, 73, 76, 80, 81, 99, 100, 103, 104, 110, 113, 114, 115, 134, 135, 138, 145, 151, 172, 178, 199, 201, 202, 203, 209
Quine, Willard Van Orman, 17

rationality, xi, 40, 41, 71, 72, 73, 85; Enlightenment, 16, 80; formal, vii; instrumental or technological, viii, x, 71, 74, 76, 80, 181, 192
rationalization, xi, 16, 49, 77, 80, 131, 133, 147, 150; alternative vii, 19, 80, 114; secondary, 112; subversive, xvi, 30, 37, 39, 40, 47
Rawls, John, 92, 99, 176, 197
reproductive technologies:, 22, 34, 39, 189, 190. *See also* technology.
Re-reading the Canon:, Feminist Interpretations of Heidegger (book), 50
Research in Philosophy and Technology (book), 110

restoration ecology, xviii, 148, 149, 150, 203
Rethinking Ecofeminist Politics (book), 51
Reweaving the World:, The Emergence of Ecofeminism (book), 52
Rhode, Deborah, 51
Rickover, Hyman, 157
Rockmore, Tom, 52
Rohm and Haas Chemical Company, 166, 172
Rosaauers Supermarkets, 106
Roth, Guenther, xix
Rowland, Robyn, 50
Ruether, Rosemary Radford, 47, 51
Russia, 157

Schaffer, Simon, 18
Schmid, A. Allan, 112, 134
Scholars and Entrepreneurs:, The Universities in Crisis, 36
Schomberg, R., 36
science, 15, 37, 41, 47, 72, 79, 80
Science (journal), 18
Science and Technology Studies (STS): 3, 6, 59, 68, 117, 131, 154, 159, 160, 167, 169, 172, 184, 186, 202
Science in Action:, How to Follow Scientists and Engineers through Society (book), 134
scientific Marxism, ix, xx
Scientific Practice:, Theories and Stories of Doing Physics (book), 18
Science Studies (or technology studies). *See* Science and Technology Studies (STS)
Science, Technology, & Human Values (journal), 110, 152, 172, 208
Sclove, Richard, 92, 93, 147, 154, 171
SCOT. *See* Social Construction of Technology.
Searle, John R., 18
Second Sex, The (book), 41, 51
Shaping Technology/Building Society:, Studies in Sociotechnical Change (book), xx, 209
Shapin, Steven, 18
Shapiro, I., 99
Shapiro, Jeremy J., xix, 16
Sharp, G., 35, 36
Shiva, Vandana, 47, 52
Sierra Club, 140, 141, 145
Simondon, 104
Sinsheimer, Peter, 172
Smith, Adam, 115
Sober, Elliott, 17
Social Construction of Technological Systems, The (book), xx, 16, 209
social constructivism (or social constructivists, or social construction of technology), xvi, xx, 5, 6, 7, 8, 10, 42, 54, 73, 74, 113, 167, 168, 184, 185, 187, 188, 191. *See also* Social Construction of Technology (SCOT), *and* constructivism.,
Social Construction of Technology (SCOT), xi, xx, 167
Social Dimensions of Science (book), 17
social ecology (or social ecologists), 141, 143. *See also* deep ecology.
Social Ecology after Bookchin (book), 151, 152
social historians of technology, 3, 6, 7
socialism, xviii, 138, 153, 155, 181
Social Problems (journal), 17
Social Studies of Science (journal), 172, 208
Social Shaping of Technology, The:, How the Refrigerator Got its Hum, xx
Social Transformation of American Medicine, The (book), 135
Southern Journal of Philosophy, The, 66
Soviet Union, 157
Spain, 106
specific intellectual(s) or practice(s), 29–32, 189
Spinosa, Charles, 67, 68, 69, 70
Sporn, Phillip, 171
Stallman, Richard, 108
Stambaugh, Joan, 51
Starr, Paul, 135
Star Wars, 29
Staying Alive:, Women, Ecology, and Development (book), 52
stem-cell research, 21, 26, 33, 34
Stevens, W.K., 152
Straus, Neil, 111
Strong, D., 134, 135, 152

Strong Programme, xx, 6
Stucke, John, 110
Studies in History and Philosophy of Science (journal), 17
STS. *See* Science and Technology Studies
Stump. David J., v, xv, 3, 17, 18, 185, 187, 188, 190, 192, 203, 209, 214
subjectivity, 44, 45
sustainable communities, xviii, 203
Synthetic Planet:, Chemical Politics and the Hazards of Modern Life (book), 171

Tatum, Jesse, 105, 110, 144
Taylor, B.P., 209
Techné (journal), 208
technocracy (or technocratic system), 43, 44, 46, 49, 50, 61, 78, 87, 90, 106, 155, 180, 181, 182, 183, 185, 192
technology/technologies:, and alienation, xvii, 116, 117, 119–122; alternative 8, 118, 142; ambivalence of xi,19, 20, 22, 35, 49, 107, 115, 119, 120, 121, 177; autonomy (or autonomization) of xiv, 4, 9, 11, 12, 13, 73, 188; and class, 26; and consciousness, viii, xiv; and culture, 26; and decontextualization, xiii, 9, 22, 33, 192; and democracy, v, vi, vii, xvii, 26, 30, 38, 44, 50, 55, 61, 62, 78, 80, 85–100, 104, 105, 106, 107, 108, 109, 136–152, 157, 168, 177, 182, 183, 185, 196, 200, 201; and determinism, 4, 13, 23, 27, 28, 41, 74, 77, 78, 103, 119, 130, 131, 142, 156, 177; and emancipation (or liberation), 21, 22, 27, 33, 34, 71, 98, 205; and embodiment, 21, 22, 27, 28, 34, 189; and essentialism, xiv, xx, 5, 8, 9, 12, 13, 14, 15, 20, 23, 53, 54, 55, 56, 57, 59, 62, 64, 65, 72, 73, 85, 87, 88, 90, 91, 185, 186, 187, 189; and finitude, 179–180; and gender, 191; and innovation, 11, 118, 121, 122, 125, 127, 153, 161, 162, 167, 170; and modernization, xiii, 3, 138; and pessimism, 41; and population, vi, 136–152; and private property, 90–91, 94 (*see also* Locke); and the posthuman (or posthumanism), 27 (*see also* Feenberg); and rationality or rationalism, viii, xii, x, 4, 13, 42, 71, 72, 80, 85, 86, 91, 147, 192; and reflexivity, 21; and social hierarchy, xii; barrier-free, 94, 95, 98; commodification of, xvii, 109, 111, 112–135, 201–202; constructive theories of, 29, 42, 54, 167, 187, 188, 191 (*see also* constructivism, *and* social construction of technology); critical theory of, v, x, xvii, 1,, 53, 65, 74, 77, 80, 119, 131, 134, 178, 191, 193, 194, 208; democratic rationalization of, xii, xiii, xvii, 80, 89, 147, 150, 199; design of, xi, xv, xvi, xvii, 20, 24, 25, 42, 61, 62, 74, 86, 91, 96, 98, 113, 177, 191, 193, 202, 203; development of, 4, 7, 9, 10, 11, 12, 13, 14, 15, 28, 48, 79, 177, 187, 188, 199, 201; distancing effect of, xiv; governance of, xviii, 153, 154, 170; essence of, xi, 12, 46, 54, 55, 56, 57, 58, 59, 62, 64, 67, 73, 185, 186, 188, 189, 192, 193, 194; history of 12, 15, 16, 57; invention of, 11; malleability of, vi, 153–170, 203; momentum of, 11, 12, 14, 188; philosophy (or philosophers) of 3, 4, 5, 12, 15, 16, 53, 54, 65, 80, 85, 130, 131, 134, 144, 180, 184, 185, 193, 195, 196, 198, 199, 202, 206, 207; politics of, x, xvi, 20, 21, 29, 33, 86, 87, 97, 103, 139, 177, 189, 196–198; reform of, xv, 73, 87, 102, 103, 113, 114, 131, 139, 142, 144, 146, 148, 149, 150, 151, 200; social reconstruction of, vi, 54, 153–170, 180; substantive theories of, xvi, 8, 23, 24, 29, 54, 55, 56, 59, 61, 65, 67, 72, 76, 186, 187, 188, 189, 190, 196; system(s) of (or as a system), 3, 10, 11, 12, 13, 14, 38, 41, 74, 113, 114, 180, 183, 184, 188, 190; transfer of, 11; transformation of v, xv, xvi, 83, 117
Technoculture and Critical Theory:, In the Service of the Machine? (book), 35
technological society, viii, 180. *See also* technology.

Technology and the Character of Contemporary Life (book), 134, 209
Technology and the Good Life? (book), 134, 135, 152
Technology and the Politics of Knowledge (book), 50, 68, 80, 99, 134, 171, 209
technology studies. *See* Science and Technology Studies (STS).
technoscience(s), 31, 32, 72, 73, 74, 75, 76, 167
Teletel System. *See* French Teletel System.
Teller, Edward, 171
Tenner, Edward, 182, 209
Terminator Seed, 124
Theoretical Perspectives on Sexual Difference (book), 51
Thompson, E. P., 121, 134, 187, 200
Thompson, Paul B., vi, xvii, 109, 111, 112, 135, 199, 201, 202, 209, 214
Thomson, Iain, v, xvi, 16, 53, 192, 193, 194, 214
Thornton, Joe, 172
Theory of Communicative Action, The (book), xix
Third World, the, 25
Torvalds, Linus, 108
Toward a Rational Society:, Student Protest, Science, and Politics (book), xix, 16
Toxic Release Inventory, 164
Transforming Technology:, A Critical Theory Revisited (book), 35, 36, 171, 177, 178, 189, 201, 203, 209
Transparency of Evil, The (book), 66
Tronto, Joan, 45, 51

U.K. *See* United Kingdom.
Understanding Computers and Cognition:, A New Foundation for Design (book), 68
United Airlines, 106
United Kingdom (U.K.) (or England or Britain), 121, 122, 157, 176
United States Federal Reserve, 163
United States General Accounting Office, 164
United States Food and Drug Administration (FDA), xiii
United States House of Representatives, 169
United States of America (U.S.), 15, 38, 136, 141, 148, 149, 157, 164
U.S. *See* United States of America.
US Airways, 106
U.S. House of Representatives. *See* United States House of Representatives.

Values at Work:, Employee Participation Meets Market Pressure at Mondragon (book), 110
Veak, Tyler, iii, 110, 208
Vietnam, 176
Vorträge und Aufsätze (book), 50

Wajcman, Judy, xx
Warner, John C., 171
Warren, Karen, 37, 47, 48, 52
Washington Post, The (newspaper), 35
Weber, Max, vii, viii, x, xi, xiii, xix, 3, 4, 6, 8, 9, 10, 11, 18, 55
We Have Never Been Modern (book), 18, 134
Western Behavioral Sciences Institute, 176
West, the, 24, 47, 57
Westinghouse, 156, 157
Whale and the Reactor, The (book), 110, 135
Whitehead, Alfred Noth, 76
Whitney, Eli, 125
Why Things Bite Back:, Technology and the Revenge of Unintended Consequences (book), 209
Wilderness Society, 140
Wills, Garry, 119, 134
Winner, Langdon, 75, 110, 112, 131, 132, 135, 154, 155, 170, 171, 198
Winnograd, Terry, 67
Wittich, Claus, xix
Woman and Nature:, The Roaring Inside Her (book), 52
Women and Values:, Readings in Recent Feminist Philosophy (book), 51
Wong, Edward, 110
Woodhouse, Edward J., vi, xviii, 153, 171, 172, 173, 199, 203, 204, 214

World War II, 202
World Wildlife Fund, 169

Yearning:, Race, Gender, and Cultural Politics (book), 52
Young, Julian, 67, 69

Zammito, John H., 17
Zen, 175
Zimmerman, Michael, 37, 50